Laboratory Inquiry in Chemistry

Second Edition

Richard C. Bauer
Arizona State University

James P. Birk
Arizona State University

Douglas J. Sawyer
Scottsdale Community College

THOMSON
BROOKS/COLE

Australia • Canada • Mexico • Singapore • Spain • United Kingdom • United States

COPYRIGHT © 2005 Brooks/Cole, a division of Thomson Learning, Inc. Thomson Learning™ is a trademark used herein under license.

ALL RIGHTS RESERVED. No part of this work covered by the copyright hereon may be reproduced or used in any form or by any means—graphic, electronic, or mechanical, including but not limited to photocopying, recording, taping, Web distribution, information networks, or information storage and retrieval systems—without the written permission of the publisher.

Printed in the United States of America
1 2 3 4 5 6 7 08 07 06 05 04

Printer: West Group

ISBN: 0-534-42424-4

For more information about our products, contact us at:
Thomson Learning Academic Resource Center
1-800-423-0563

For permission to use material from this text, contact us by:
Phone: 1-800-730-2214
Fax: 1-800-730-2215
Web: http://www.thomsonrights.com

Brooks/Cole—Thomson Learning, Inc.
10 Davis Drive
Belmont, CA 94002-3098
USA

Asia
Thomson Learning
5 Shenton Way #01-01
UIC Building
Singapore 068808

Australia/New Zealand
Thomson Learning
102 Dodds Street
Southbank, Victoria 3006
Australia

Canada
Nelson
1120 Birchmount Road
Toronto, Ontario M1K 5G4
Canada

Europe/Middle East/South Africa
Thomson Learning
High Holborn House
50/51 Bedford Row
London WC1R 4LR
United Kingdom

Latin America
Thomson Learning
Seneca, 53
Colonia Polanco
11560 Mexico D.F.
Mexico

Spain/Portugal
Paraninfo
Calle/Magallanes, 25
28015 Madrid, Spain

Table of Contents

Acknowledgments iv
Preface v

Your Role in Learning... 1
Your Group.. 1
Your Laboratory Instructor 2
Proposals and Class Discussions 2
Laboratory Notebooks 3
Scientific Reports 4

The Investigations

1. What Are the Safety Concerns in the Laboratory?... 6
2. What's in the Flask? 21
3. How Should Data Trends Be Presented? 25
4. How Is Lab Equipment Used? 29
5. What's in the Bottles? 33
6. How Can the Waste Be Made Useful? 37
7. Is the Water Hard or Soft? 41
8. How Hot Is the Water?.. 45
9. Which Metal Will Burn the Skin? 49
10. Are All Neutralization Reactions the Same? 53
11. How Much Sodium Bicarbonate Is in the Mixture?.. 57
12. Is It Economical to Recycle Aluminum? 61
13. What Is a Copper Cycle? 65
14. Who Wrote the Ransom Note? 69
15. How Can UV Sensitive Beads Be Used to Test Sunscreens? 73
16. What Factors Affect the Intensity of Color? 77
17. How Much Cobalt Is in the Soil? 81
18. How Much Copper Is in the Coin? 85
19. Which Iron Compound Is It? 89
20. Should We Mine This Ore? 93
21. What Causes Intermolecular Attractions? 97
22. What Are the Structures of Some Alloys? 101
23. How Is LED Light Color Related to Composition? 105
24. What Is the Molar Mass of Mars Ice Gas? 109
25. Are Pollutant Gases Harmful to Plant Life? 113
26. How Much Gas Is Produced? 117
27. Which Alcohols Are in the Barrels? 121
28. How Is Heat of Combustion Measured Indirectly?... 125
29. What Is the Rate Law? 129
30. How Fast Does the Crystal Violet Decolorize? 133

31. Why Is the Vinegar Factory Rusting? 139
32. What Factors Affect the Solubility of Kidney Stones? 143
33. How Many Chemicals Are in the Vial? 147
34. What Factors Affect Chemical Equilibrium? 151
35. What Is the Formation Constant? 155
36. Are Household Items Acidic, Basic, or Neutral? 161
37. What Is the pH of Soil? 165
38. What Is the Acid Dissociation Constant? 169
39. What Is the Solubility Product? 173
40. What Are Some Chemical Properties of Cream of Tartar? 177
41. What Are the Metals? 181
42. How Can a Battery Be Made from Coins? 185
43. What Is the Complex Ion? 189
44. What Formulation Makes the Best Toy? 193
45. How Are Anions Identified? 197
46. How Are Cations Identified? 203
47. How Are More Cations Identified? 207
48. How Are Ionic Solids Identified? 211
Presentations and Poster Sessions 215

Appendices

A. Automated Data Collection 219
B. Transmittance and Absorbance Data Collection 221
C. Measuring pH 224
D. Temperature Data Collection 227
E. Pressure Data Collection 228
F. Voltage Data Collection 231
G. Selected Laboratory Techniques 232
H. Laboratory Equipment 244
I. Sample Material Safety Data Sheet 245
J. Tables 251

Acknowledgments

The development of this manual involved the hard work of many people at Arizona State University and Scottsdale Community College. We would not have completed this book were it not for the contributions made by the students who piloted these experiments. We also want to thank the lab instructors for piloting these unconventional lab exercises, sometimes with little guidance about what to expect from students left on their own to design experiments. Support staff at both institutions, in addition to their regular duties, helped in the preparation of student and instructor materials during the piloting stage of development. Finally, we would like to thank the editorial staff at Brooks/Cole for its help in developing this work.

Preface

Chemistry is a discipline in which scientists inquire about the microscopic and macroscopic worlds around us to understand the chemical nature of our surroundings. The basis for this inquiry is experimentation in which chemists probe for answers to scientific questions we face in our world. Sometimes, chemists explore the microscopic world with well–established experimental techniques. More often, however, they must design their own experiments, adapting techniques to their specific problems. We hope that this laboratory experience helps you to develop important problem solving skills necessary for success in our competitive, technological society. With guidance from your laboratory instructor, the techniques described in this book, and your text book, you are free to use your creativity in solving the problems we pose here.

The inquiry-based investigations in this book are designed to foster your experimental problem solving skills as well as to help you learn laboratory techniques. Because of the experimental context, some things you do in the laboratory may not coincide directly with your lecture discussions. Some of the ideas you will confront in lab will not be examined in lecture because they are issues appropriate only for the laboratory. For example, we hope that this lab experience helps you develop experimental design skills. These skills are not easily developed in a lecture setting. While experimental skills depend on conceptual knowledge, they differ from the concepts you will learn in lecture. Some of the investigations will ask you to draw on conceptual information that may seem unfamiliar. Don't panic when you are confronted with these concepts. The textbook for the lecture component of this class will be a valuable resource in helping you relate the unfamiliar material to the problem at hand. Practicing scientists must often consult textual resources and colleagues to solve their problems. They often encounter difficulties, solutions for which are not neatly outlined in detailed laboratory procedures. Instead, they design their own experiments with the knowledge of experimental techniques in mind. Hopefully, this lab experience will expose you to the critical thinking skills used by practicing chemists; that is, we hope that from the lab experience you can "learn to learn," a skill demanded by employers of college graduates in any discipline.

Scientists and, indeed, members of nearly all other disciplines work in a community with others having similar interests. Employers are demanding that undergraduate education prepare future employees for working in a cooperative, team environment. In a spirit of collegiality, we have designed the laboratory investigations to maximize the contributions others can make in improving your understanding of chemistry and to prepare you for a workforce that depends on cooperative skills. These skills include communicating ideas, negotiating with others, valuing the contribution of all group members, and delegating tasks. During your laboratory meetings, you will draw from the understanding of your fellow students to help your group design and carry out the investigations. The group is responsible for the success of each individual in it.

In the laboratory you will be asked to draw on your knowledge of chemical concepts as well as your interpretation of appropriate experimental procedures. Because science is often a creative endeavor, we have removed the "recipes" that often accompany laboratory experiments in general chemistry. Planning investigations is often neglected in typical recipe experiments, where the design aspects are done by the experts who wrote the lab manual. These experts learned their experimental skills by trial, error, creativity, serendipity, and knowledge of chemical behavior. By removing the recipes we hope to encourage you to develop these critical skills. You are responsible, together with your fellow group members, for the overall planning of the experiment which includes the design, data collection, organization, interpretation, and communication of your laboratory investigations. This accountability will at times be frustrating because you may not feel intellectually equipped to complete the investigation. However, once you draw on your understanding of the world, the chemistry you have learned so far, and the experience of others, we are confident that you will find the investigations rewarding.

Richard C. Bauer
Department of Chemistry and Biochemistry
Arizona State University
PO Box 871604
Tempe, AZ 85287-1604
rbauer@asu.edu

James P. Birk
Department of Chemistry and Biochemistry
Arizona State University
PO Box 871604
Tempe, AZ 85287-1604
jbirk@asu.edu

Douglas J. Sawyer
Department of Chemistry
Scottsdale Community College
9000 E. Chaparral Rd.
Scottsdale, AZ 85256
doug.sawyer@sccmail.maricopa.edu

Your Role in Learning

We have found that successful chemistry students do some chemistry every day. Does a college basketball team practice once a week for ten hours? No, the team practices several times a week for shorter durations. You should approach your study of chemistry in the same manner; that is, practice chemistry every day for a little while instead of doing marathon cramming sessions once a week. The laboratory provides you with an opportunity to practice your chemical understanding in class while you plan and conduct your investigations with your team members. Indeed, your laboratory success depends on your understanding of chemical concepts, your ability to regularly put your understanding into practice, and your capacity to communicate your ideas orally and in writing. Don't be afraid to voice your ideas during your lab meetings and when your group meets outside of class. Putting your thoughts into words is a critical exercise in helping you develop cognitive skills. Also, your fellow students and your instructor can help you make better sense of the material provided they know what you understand about the subject at hand.

As stated previously, we hope the lab environment helps you to draw on the experience of others in the course. Hopefully, your fellow group members will help you succeed in this course. Likewise, you are responsible for the success of each team member. You must take your role in your team seriously for you and your fellow group members to succeed. Just as individuals are accountable for their work in the "real" world, so too will you be held personally responsible for gaining an understanding of laboratory procedure and chemistry concepts.

As in other disciplines, learning chemistry requires your active immersion in the discipline. No matter how good (or bad) your instructor or lab partners are, your personal commitment to the course is the key to your success in chemistry.

Your Group

The organizational structure of most companies is team based. Employers have found that in a team-structured environment, gains are made by people putting their heads together to solve problems. In educational settings, group-based, cooperative learning efforts improve relations across different cultures and genders, increase student retention in difficult science courses, and generally improve students' performance. In industrial, governmental, and educational institutions, group work often has at least four characteristics: (1) the group works toward a common goal, (2) the group is responsible for the success of each member, (3) each member is individually accountable for completing various tasks and gaining relevant knowledge, and (4) the group members share resources. At the beginning of the course your instructor will assign you to a group in which you will conduct the investigations throughout the semester. You will act as a team of investigators to solve the assigned problems posed in this book. Socially, you may not have anything in common with other members of your group. However, successful team efforts often rest on the ability to draw from the experiences of people from diverse backgrounds.

To facilitate your group's success, each member could assume one of the following roles:

Team leader — coordinates the activities of group members, including the generation of reports.

Assistant leader — keeps the group on task. Consolidates supporting literature relevant to the investigation. Monitors group activities so all members contribute in a healthy, supportive environment.

Data collection expert — directs the data collection activities for the group. This includes identifying the information that must be collected and recording the data in an appropriately organized manner.

Experimental technique expert — determines the appropriate reagents and equipment assembly for data collection. Consults with instructor for special instructions about equipment.

After each investigation these duties should rotate so each group member has an opportunity to develop the associated skills of each role. Even though you will assume various roles, each group member is expected to contribute to experimental design discussions. Your instructor will monitor your group's activities so all members take part.

Because each member of your group is individually accountable for successful completion of the class assignments, you will be asked to describe your contribution to each investigation. At the end of each investigation is a contribution form that must be completed by every member of the group. On that form you will report your percent effort as well as your perception of what other group members contributed. In addition, you will also describe what you did to help your group complete the investigation and write the report. The contribution forms are due at the same time as your report.

Your Laboratory Instructor

In light of our goal to promote creative student development of experimental design and technique, your laboratory instructor will serve as a resource, not as a dictator of "correct" problem solutions. Do not expect your instructor to tell you exactly what to do. Your success in the course depends on your ability to think for yourself. Too much input from your instructor will undermine the learning process during your laboratory investigations. Use all the resources at your disposal before you approach your instructor with questions. These resources include the lab book, your textbook, your fellow students, the Internet, and your library. However, when you are uncertain of specific techniques or unfamiliar with laboratory equipment, consult your instructor before carrying out your experiment. Once your experimental design is in order, you **must** get your instructor's approval before proceeding with your investigation.

Proposals and Class Discussions

Many of the investigations in this manual are designed to be completed in two weeks. The first week of an investigation will involve designing your experiment to solve the problems posed. This is not a trivial task, so you can expect it to take some time. You must consider equipment issues, determine which reagents and quantities you will use, and decide on variables to control, data to collect, and how you will analyze the data. In addition, you must also address safety considerations. Once your group has decided on an experimental design, you will need to present your ideas to your instructor. Your instructor may require a written proposal before your actual experimentation begins. Your proposal should address how the experiment will yield a solution to the problem. You should describe the data you will collect and materials and quantities you will use. Finally, you must describe the safety hazards associated with your experiment and appropriate precautions you will take to avoid personal injury. There may be times when you want to begin by running simple experiments to make observations or by collecting preliminary data before you decide on your experimental design. It is perfectly acceptable to run a simple initial experiment but you must obtain the instructor's approval before you proceed.

At various points during the investigation your instructor may ask each group of students to give a short progress report orally to the rest of the class, much as practicing chemists do in scientific meetings. These reports may occur midway through an investigation or at the end. Part of your grade will be based on the clarity and creativity of your presentation.

Laboratory Notebooks

Scientists maintain laboratory notebooks as a permanent record of their laboratory activities. They always refer back to them when, for example, they apply for patents or compose scientific papers. Additionally, lab notebooks are often used in litigation when scientific laboratories are taken to court. Because of these important implications it is imperative to learn data collection and recording techniques that prepare you for future scientific study. Careful recording of experimental observations and results will be encouraged and enforced. You should develop a style of notebook that is both convenient for yourself and intelligible to others. There are some features of the format that are dictated by universal convention, while others are largely a matter of personal taste. There is an inevitable tendency to make temporary records or to trust your memory, presumably to make the final copy look neater, but this is a bad habit which you should avoid. Those aspects of the format for recording data that must be observed are all related to the idea that the notebook should be a permanent, documented, and primary record of laboratory observations. Use the following guidelines in maintaining your lab notebooks:

1. Research data are kept in bound notebooks with prenumbered pages. Because of limited space at your lab station, your instructor may permit you to have a spiral bound lab notebook. The notebook should be dedicated to a single subject. Your instructor may want your notebook to contain carbonless duplicates that can be torn out and turned in.

2. The lab notebook will serve as a permanent record of what you did, data you collected, and observations you made in the laboratory. Someone wanting to duplicate your experiment should be able to do so using your notebook. You should also include calculations and summaries.

3. All entries should be written permanently in ink. Mark out mistakes with a single line.

4. Sign and date each page.

5. Leave the first page blank to include a table of contents.

6. In lieu of rewriting a lab procedure, it is acceptable to cut and paste a copy of the procedure into the lab notebook with appropriate references.

7. Some of the data you collect may be processed with a computer. You may paste the printout of the data in your lab notebook. However, in your lab notebooks you must still make a written record of the data you collect and observations you make. Your instructor will occasionally inspect your data and observations from your lab notebooks.

8. While your group will have a data collection expert for each investigation, each student must maintain a record of data collected. The manner in which you obtain data from your group is at your discretion.

Scientific Reports

The scientific report is one manner in which investigators communicate results of their experiments. They are usually written in impersonal or third person, passive voice. You will be required to write laboratory reports about your work to give you practice in transmitting scientific results to others. A report should be brief and to the point, and should be readable, both grammatically and stylistically. Reports must be typed. Your report will be judged on the quality of your work as well as on its presentation, its interpretation, and the reasoning behind the interpretation. While the exact format of the report is up to you, the report should contain the items listed below (assuming they are appropriate to the experiment). The format given below is that generally used for the reporting of scientific results.

Introduction: This should contain a clear statement of the problem, its goals, and your general approach to solving the problem. A typical introduction might be a short paragraph in length. As you do more original work, the significance of the present work in the context of what is known should be stated. Most scientific papers include a discussion of background theory in the introduction. Your instructor may direct you to include this information. Do not copy the goals listed for the investigations in your report introduction. Those goals merely represent a checklist of items to complete during the investigation. The opening question in the investigation title should provide a guide fro writing you introduction.

Experimental: Enough detail should be given in this section so that someone else, not otherwise familiar with the work, could repeat the experiments. Do not write your experimental section in second person voice. You are describing what you did, not providing directions for someone else to follow.

(a) You must identify all materials used. Include any information you have on the purity and concentrations of the materials. You should list reagents, such as acids, bases, and solvents, that are normally available, but you need not describe them in detail. You should give formulas of all compounds, together with their chemical names, at least once.

(b) Unless it is standard equipment, you should describe apparatus, with a drawing if necessary, and with names and sources of specific equipment if they are not widely known. You should list commercially available equipment, but you need not describe it in detail.

(c) Include a description of how the work was done. For experiments involving established procedures, reference to the appropriate source of the procedure may suffice. You should include all the background data, equations, and formulas necessary to the experiment.

Results: For most reports, the presentation of the results and the discussion of their significance may be separated into two distinct sections. Occasionally, however, a chronological approach might be preferable. Regardless of the outline you follow, several points should be noted:

(a) The major experimental results, including the original data, the calculated results, and at least one detailed sample calculation showing how the final results were obtained should be presented. It may be appropriate to include the theory behind your calculations. Include only relevant data and describe assumptions you made in the collection of the data. Introduce equations, figures, graphs, and tables where necessary for clarity and conciseness.

(b) All numerical data should be reported in accepted, self-consistent systems of units. Report the precision of the work, theoretical values if known, and the relative error of the experimental result.

Discussion (or Conclusions): You should discuss your results in detail.

(a) In the discussion of the significance of the results, an objective explanation is essential. You should point out the limitations of the work. You should also interpret, compare, and contrast your results with reports available from other sources. Background theory should be used to explain your results. Try to correlate your results with the chemical principles or reactions involved.

(b) If your results differ significantly from expected values, or if the precision is worse than should be obtained with your procedure, discuss the possible sources of error in detail.

(c) A summary adds to the value of the presentation. It should be interpretive and not repetitious. Discuss how the results relate to the goals of the experiment and any conclusions that may be drawn from the experiment. Consider any alternate conclusions or explanations. The problem may not have been solved completely; if so, you might suggest an approach or a refinement that could be used for further study.

Investigation 1
What Are the Safety Concerns in the Laboratory?

Introduction

Working in a chemistry laboratory has certain inherent dangers. You can avoid laboratory hazards with the proper knowledge of the materials and chemicals with which you will be working. In this investigation you will acquaint yourself with safety rules implemented for the protection of everyone in the lab. In the event that a dangerous situation does arise, this investigation provides guidelines for dealing with these occurrences. In addition to answering the questions on the worksheet, you will identify the safety rules that your team believes are the most important.

Goals

As you complete this investigation you will:
1. Become familiar with rules designed to maintain a safe laboratory environment.
2. Become familiar with procedures for handling chemicals.
3. Select the safety rules that you think are most important and must be followed at all times.
4. Informally summarize your chosen safety rules in class discussion and formally write up your conclusions in a report.

Materials

Reading materials in this investigation
Other materials provided by your instructor

Getting Started

Based on the materials provided, and a tour of the lab, work with your group to develop answers to all of the questions on the worksheet at the end of this investigation. If you are uncertain of an answer, check it with your instructor. Work with your group to select the five rules that you think are most important to ensure your safety during laboratory sessions. Be prepared to discuss at least one of your chosen safety rules with the rest of your class.

Report

As a group, review the safety rules and answer the questions on the worksheet. Each student must submit answers to the worksheet questions at the end of the lab period. In addition, your group must submit a report justifying your team's five most important safety rules. Your written report should include a one-paragraph justification for each of your choices. The report must be typed and grammatically correct. It will be returned to you for corrections if it is not acceptable. Each member of the group should be prepared to discuss your selections with the rest of the class.

Safety Rules and Laboratory Procedures

For many of the investigations you will complete in this lab class, your instructor will relate specific safety information regarding the chemicals or procedures you will use. Instructors will do their best to provide a safe working environment for all the students. However, you have a responsibility to everyone in the lab to understand and follow the safety rules listed below. For everyone's protection, your instructor will strictly enforce the safety rules. Failure to abide by them may result in your dismissal from the class.

- *Plan your investigations in advance.* Many lab accidents occur because the experiment is poorly planned or disorganized. A well-planned experiment not only reduces accidental risk, but will also ensure that your data collection will proceed in a smooth manner.

- *Experiments must be approved by your instructor.* While we encourage creativity in solving the problems we pose in this manual, your instructor must approve your experimental designs. Each investigation requires a formal proposal to your instructor before you can proceed with your experiments. (See page 3 for a discussion of lab proposals.) Your instructor will approve your proposal after careful consideration and discussion of safety implications. If you choose to modify a common procedure, your modification must also be approved by your instructor.

- *Never work in the laboratory alone or without your instructor present.* If an accident occurs an authority must be present to hasten medical attention. Also, you should always be present while your experiments are in progress.

- *Food and beverages are not allowed in the laboratory.* Even if you plan to consume them outside of lab, you run the risk of chemical contamination. This applies to chewing gum and chewing tobacco, as well. You should store your personal effects away from your work area.

- *Smoking in the laboratory is prohibited*, for obvious reasons. Flammable solvents are often present in the lab and can easily explode or ignite in the presence of a flame. You also run the risk of chemical contamination of your cigarettes (as if they don't already have enough toxic chemicals in them).

- *Immediately report all injuries, including minor ones, to your instructor.* The presence of chemicals can complicate even minor cuts.

- *Recommendation: Get health insurance.* We recommend that anyone enrolled in a chemistry course obtain personal health insurance. Enrollment in courses does not automatically entitle you to insurance coverage if an accident occurs in the lab. If your campus has a health center it may be capable of treating some injuries. However, further medical attention may require your transport to other facilities. You are responsible for the cost of those additional services.

Investigation 1

Safety Equipment and Laboratory Attire

The prevention of accidents for you personal safety should be your first priority when you conduct laboratory experiments. Your instructor will strictly enforce safety rules regarding appropriate safety equipment and attire.

- *You must wear safety goggles at all times while experimental work is in progress in the laboratory*, even if you have completed your experimentation. Many injuries occur because of flying debris or splashed chemicals originating some distance from the victim. Safety glasses are unacceptable for eye protection in a chemistry laboratory. In addition, you should not wear contact lenses in a chemical laboratory. Chemicals can splash under the lens and cause permanent damage to the cornea. Furthermore, noxious gases can permeate soft lenses.

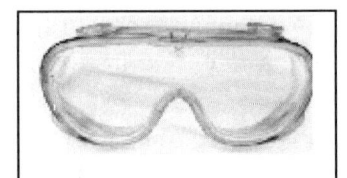

- *Make every effort to protect your bare skin.* Minimally, you should wear long pants, shirts with sleeves, and shoes. We also advise you to wear a plastic apron or a lab coat to protect you and your clothing from corrosive chemicals. Restrain loose clothing, hair, and jewelry to prevent injury. For protection against broken glass or spilled chemicals, *wear shoes at all times in the laboratory.* Sandals are not appropriate footwear for the laboratory.

- *Identify the location of the safety equipment such as safety showers, eyewashes, fire extinguishers, first-aid kits, respiratory equipment, and emergency exits.* Your instructor will call your attention to these items during your first laboratory period. You should understand how safety equipment is used.

Cleanliness

- *Keep your assigned bench area neat and free of spilled chemicals.* A messy bench is a potential safety hazard. An unorganized bench also leads to confusion that prohibits the efficient running of the investigation. Experiments will proceed more rapidly and safely in a neatly kept area. When you enter the laboratory it should be clean. If not, you must clean your area before you proceed. You must also leave the lab clean when you depart.

- *Spilled chemicals must be cleaned up immediately.* Occasional spillage is unavoidable and will cause no damage if it is immediately sponged off with water. However, many solutions and solids will attack the bench sides and tops if allowed to remain on them for any length of time. Spills can also cause chemical burns if you accidentally rub against them. If you spill a large volume of water or other liquid it must also be cleaned up immediately. If the material is corrosive or flammable, ask the instructor for assistance. If acids or bases are spilled on the floor or bench, neutralize with solid sodium bicarbonate, then dilute with water. Mercury is very toxic, even in small quantities (for example, from broken thermometers). Notify the instructor immediately in case of any spillage of mercury.

- *You must dispose of all laboratory waste in appropriate containers.* EPA regulations require that all chemical waste be disposed of properly. Failure to comply with EPA standards has resulted in heavy fines leveled on individuals as well as institutions. In planning your experiments, keep in the mind the waste implications of your selected design. Waste bottles for each experiment will be provided by your instructor. They will be labeled according to the specific contents that should be placed inside. *Flammable liquids should be disposed of in specially marked bottles*, never in the sink. Do not place any materials in the sink since damaging overflows may result. Keep lab sinks and troughs clean at all times. *Broken glass should be discarded in the special containers provided.* Ask your instructor for assistance if necessary.

- *Clean your bench area and common lab areas before you leave.* Make certain that the gas and water sources are turned off and that your drawer (if you have one) is closed and locked. Always use clean apparatus for each experiment. Dirty equipment is harder to clean later and will require considerable scrubbing the next time it is needed. Clean glassware by scrubbing with a brush, hot water, and detergent, then make a final rinse with a stream of distilled water. Do not dry glassware with compressed air because it is often oily.

- *Return any special equipment* which you obtained for an experiment (for example, hardware, rubber stoppers, rubber tubing, test tubes, files, volumetric flasks, and so on) at the end of the laboratory period to the location from which it was obtained. Students in other periods will probably need to use this equipment.

Laboratory Procedures

- *Balances*: Keep the balances and the surrounding area clean. *Never place chemicals directly on the balance pans.* Instead, place a piece of paper or a small container on the pan first, then weigh your material. Never weigh an object while it is hot because of the danger to you and to equipment. Also, the hot item creates air currents around the balance that can lead to inaccurate mass measurements.

- *Pipetting:* Never pipet any liquid directly by mouth! Use a rubber bulb instead. The convenience of mouth pipetting is not worth the risk of ingesting toxic chemicals.

- *Burners:* Avoid burns by lighting burners properly. After checking that the gas jet was not left open from a previous class, turn on the gas. Light the match away from the burner. Slowly move the lit match to just over the side of the barrel top, keeping your hand and fingers away from the flame area. Be careful when near the flame — especially watch loose clothing.

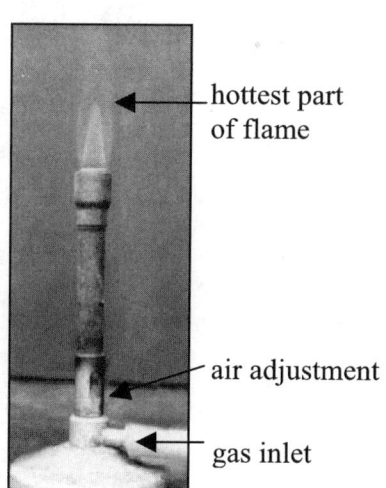

Investigation 1

- *Heating:* Always use a test tube holder when heating a substance in a test tube as shown in the picture. Be careful not to direct the tube toward yourself or anyone nearby. A suddenly formed bubble of vapor may eject the contents violently and dangerously. *Never apply the direct heat of a flame to heavy glassware* such as volumetric flasks, burets, graduated cylinders, bottles, mortars, pestles, and thermometers. These items can break spectacularly when heated strongly. Volumetric glassware also becomes distorted from the calibrated volumes. Avoid heating any object too suddenly. Apply the flame intermittently at first.

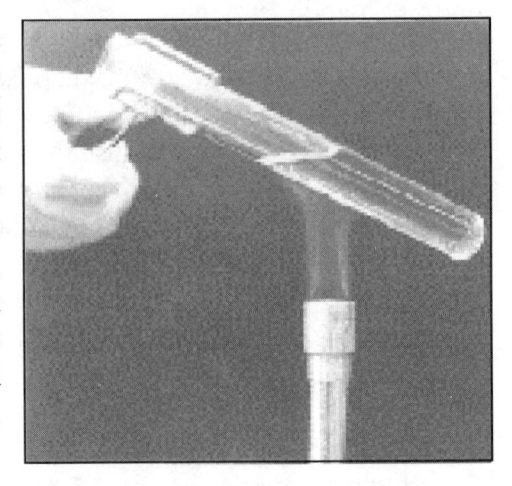

  Give heated objects adequate cooling time. During any procedure involving the heating of an object, avoid burns by thinking twice before touching anything. *Give heated glass adequate time to cool*; it looks cool long before it is safe to handle. Use the appropriate type of tongs when moving hot objects. Test tube holders should be used for test tubes only, as they are too weak to carry flasks or other heavier objects.

- *Glass tubing:* The most common laboratory accidents occur because of improper insertion of glass objects into rubber stoppers. If you intend to insert a piece of glass tubing or a thermometer into a hole in a rubber stopper, *lubricate the glass tubing* with a drop of glycerin or water, *hold the tubing in your hand close to the hole, and keep the entire assembly wrapped in a towel while applying gentle pressure with a twisting motion. Use only tubing which has been fire polished.*

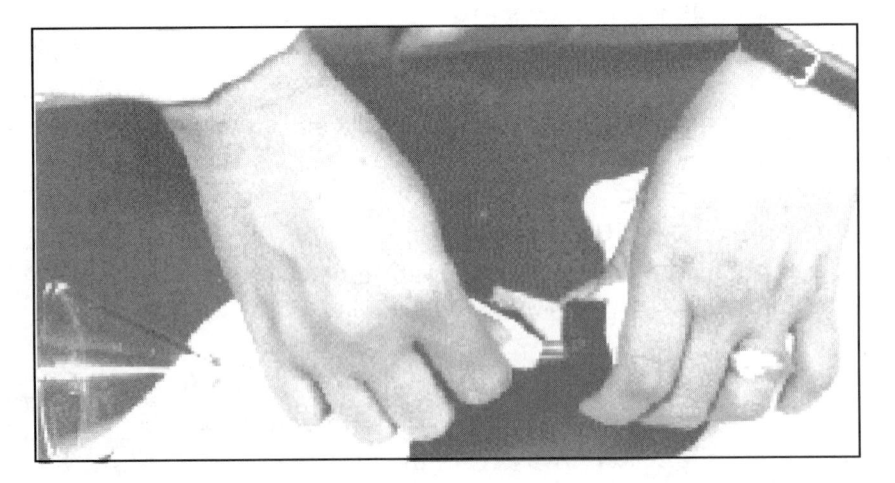

10   *What Are the Safety Concerns in the Laboratory?*

Chemicals

- *Always consult proper documentation before working with any chemical.* OSHA regulations require that Material Safety Data Sheets (MSDSs) be available to those working with chemicals. Your instructor will show you where the MSDSs are located. A sample can be found in Appendix I of this manual. There are good Web sites that have MSDSs (for example, see *http://www.fisher1.com*). At the beginning of each investigation your instructor will discuss the dangerous properties of any materials for which special precautions are necessary.

- *All chemicals should be considered as potentially toxic. Never taste a chemical or solution.* If you are asked to smell a chemical, *gently fan the vapors toward your nose.* You might ask your instructor to demonstrate the proper procedure for smelling a chemical. Nearly all chemicals are poisonous to the human body to some extent. Latex gloves are available to protect your hands from chemicals that may irritate the skin.

- *All chemicals which come in contact with the skin should be considered toxic and immediately washed off* with soap and copious quantities of water. If the area of contact is large, use the safety shower. If the eyes are splashed with a liquid, immediately flush with the eyewash. Many toxic organic compounds, which are not corrosive, are absorbed through the skin with no immediately visible symptoms. Make a habit of washing your hands before leaving the laboratory and never remove any chemicals from the laboratory. If you receive a chemical burn, after washing with water, seek medical treatment.

- *When diluting concentrated acids, always pour the acid slowly into the water with continuous stirring.* Heat is liberated in the dilution process and if water is poured into acid, steam may form with explosive violence, causing splattering.

- *Use the fume hood for all experiments involving poisonous or objectionable gases or vapors.* Never purposely allow fumes or volatile liquids to escape into the open room. Inhalation of more than a slight dose of such fumes should be reported and you should get to fresh air immediately. Also get fresh air if you are exposed to lab fumes for prolonged time periods.

- *Never use an open flame and flammable liquids at the same time.* Also, if your procedure calls for use of a flammable liquid, carry out the experiment in the hood. Do not heat anything other than water or an aqueous solution directly over an open flame. Consider all solvents other than water at least as flammable as gasoline. Many, such as ether, are more flammable.

Reagents

- *Read the label carefully before taking anything from a bottle.* Using the wrong material could result in a serious injury. *Do not take reagent bottles from the reagent shelf to your work area.* Instead use test tubes, beakers, or paper to obtain chemicals from the dispensing area. When you take chemicals from reagent bottles keep in mind the cost of the chemical and waste disposal of unused quantities. However, to avoid possible contamination *never return unused chemicals to the reagent bottles.*

- *Do not insert your own pipets, droppers, or spatulas into the reagent bottles.* In order to avoid contamination of materials, pour them from the bottles. When pouring liquids from glass-stoppered reagent bottles, grasp the stopper between two fingers of the hand with which you hold the bottle to pour. Never place the stopper on the bench, and always replace each stopper before using another bottle, in order to avoid contamination. To avoid taking excess material it is often convenient to obtain an approximate amount by dispensing the material into a beaker. For solids you might take an approximate amount by pouring the solid into the inverted bottle cap. When pouring solid chemicals, rotate the bottle to control the rate of flow from the bottle. Do not tap or pound. If chemicals are spilled on the outside of a bottle while pouring, wash the bottle off. If chemicals are spilled on the reagent bench, clean the bench immediately.

Other Safety Issues

There may be other issues associated with safety that have not been addressed in this investigation. Check with your instructor about issues specific to your locality. In case of emergency you should know the location of the nearest phone. You should also know evacuation procedures.

What Are the Safety Concerns in the Laboratory?
Worksheet

1. Describe the dangers of not wearing goggles in the lab. What conclusion about when to wear goggles can you draw from this description?

2. Describe the dangers of wearing tank tops, shorts, or sandals in the lab. What clothing items are most appropriate for laboratory work?

3. What hazards are posed by loose or dangling hair, clothing, and jewelry?

4. What are the dangers of eating, drinking, and smoking in the lab?

5. Where should you store your personal effects when they are not needed? What are the dangers of leaving them on your lab bench?

6. What are appropriate procedures for handling lab injuries?

Investigation 1

7. Describe the location of the nearest

   (a) safety shower.

   (b) eyewash fountain.

   (c) fire extinguisher.

   (d) first-aid kit.

   (e) emergency exit.

8. What should you do if chemicals are spilled on lab benches?

9. What should you do if you spill some mercury?

10. What are the consequences of drying glassware with compressed air?

11. What are proper procedures for weighing chemicals on balances?

12. Why should you always use a rubber bulb when pipetting?

13. How should you dispose of

    (a) acids and base?

    (b) aqueous solutions?

    (c) small amounts of water-soluble solids?

    (d) large amounts of water-soluble solids?

    (e) insoluble solids?

    (f) broken glass?

    (g) flammable liquids?

14. Describe the procedure for inserting glass tubing into a rubber stopper.

15. Why should you never heat heavy-walled glassware (volumetric flasks, burets, graduated cylinders, and so on) directly with a flame?

16. How should you move a hot flask? Why is a test tube holder inappropriate?

Investigation 1

17. How should you smell a chemical?

18. What should be your immediate reaction if chemicals come into contact with your skin?

19. When is it appropriate to taste a chemical or solution?

20. Should you pour acid into water or water into acid? Why?

21. When should you use the fume hood?

22. What precautions must be taken when working with flammable liquids?

23. What are the consequences of returning unused chemicals to the reagent bottles?

24. What information is generally provided on a MSDS? With a chemical of your choice, look up the proper handling procedure for that chemical.

Investigation 1

What Are the Safety Concerns in the Laboratory? **Team Contribution Form**

Identify the percent contribution each team member made in this investigation, listing your name first.

Name: Percent Contribution:

_____   _____

_____   _____

_____   _____

_____   _____

Describe your contribution to the completion of the investigation and the writing of the report.

Return this sheet to your lab instructor.

Investigation 1

What Are the Safety Concerns in the Laboratory?
Worksheet for Second Semester Courses

1. Describe the dangers of not wearing goggles in the lab. What conclusion about when to wear goggles can you draw from this description?

2. Describe the dangers of wearing tank tops, shorts, or sandals in the lab. What clothing items are most appropriate for laboratory work?

3. What are the dangers of eating, drinking, and smoking in the lab?

4. What are appropriate procedures for handling lab injuries?

5. Describe the location of the nearest

    (a) safety shower.

    (b) eyewash fountain.

    (c) fire extinguisher.

    (d) first-aid kit.

    (e) emergency exit.

6. What should you do if chemicals are spilled on lab benches?

7. How should you dispose of any chemical?

8. Why should you never heat heavy-walled glassware (volumetric flasks, burets, graduated cylinders, and so on) directly with a flame?

9. How should you smell a chemical?

10. What should be your immediate reaction if chemicals come into contact with your skin?

11. When is it appropriate to taste a chemical or solution?

12. What are the consequences of returning unused chemicals to the reagent bottles?

Investigation 2
What's in the Flask?

Introduction

Chemists often spend time designing experiments to determine the identity of unknown substances. For example, environmental chemists analyze water samples obtained from water treatment facilities. They design their experiments to determine the identity of the substances present in the liquid and the relative abundance of each component. In this experiment you will hone your investigative skills while you become acquainted with the planning stages of experimental work.

Part 1

Suppose your team of chemists has been contacted by a municipal water supplier to recommend procedures for determining the identity of an unknown liquid. A sample has been obtained and placed in the laboratory. Your task is to design experimental procedures that could characterize the unknown's purity, physical properties, and chemical behavior. All of these may ultimately help in revealing the identity of the liquid.

Part 2

In addition to designing experiments to identify the unknown, you will become acquainted with making laboratory measurements by determining the densities of cold tap water, hot water, ice, and an unknown solid.

Goals

As you complete this investigation you will:
1. Design experiments that will identify the composition of an unknown liquid.
2. Determine the densities of cold tap water, hot tap water, and ice.
3. Determine the density of an unknown solid.
4. Write a report that provides reasonable explanations for observed differences in densities.

Materials

Unknown liquid
Unknown solid
Graduated cylinders
Beakers
Lab balances
Rulers
Other miscellaneous lab equipment

Investigation 2

Getting Started

Part 1

In planning your investigation to identify the composition of the unknown liquid, you can assume that standard experimental equipment is at your disposal. You will also be able to consult with your fellow investigators. Your team will submit a proposal to your instructor that includes ways you would determine the properties of the liquid and justification for your procedures. Also, include advantages and disadvantages for your selected experiments. Your proposal must be legibly written in standard English following accepted grammatical and spelling rules. You will not actually carry out your experiments unless permission has been granted by your instructor.

Part 2

Design an experiment to determine the density of the tap water in your lab. Since density is defined as mass per unit volume you will have to measure both those quantities. You may want to consult your textbook for further information about density. Determine the density of ice and compare it to the density of liquid water. Also, determine the density of the unknown solid.

Before carrying out your density determinations, submit a brief proposal to your instructor for approval. Use the guidelines described on page 3 for generating a proposal.

Report

For *Part 1* of this investigation, turn in your group's ideas for identification of the unknown liquid in the form of a proposal as directed by your instructor. For *Part 2*, you will write a formal report about the density experiments your group ran. Your report should clearly state what you found regarding the densities of water, ice, and the unknown solid. In addition to writing your report according to the criteria described on pages 4 and 5 of this lab manual, your discussion should address the items listed below. Try to explain differences in observed densities by considering microscopic arrangements of the atoms or molecules involved in the materials you studied. In your final report you should address the following items:

- Does the density of water from the hot water tap differ from that of the cold water? If it does, explain why.
- Explain the difference between the density of the solid and the density of water.

> **Caution:**
> **While working in the laboratory wear your goggles at all times.**
> **You do not know the identity of the unknown liquid so use**
> **caution if you decide to handle the bottle.**

What's in the Flask? Team Contribution Form

Identify the percent contribution each team member made in this investigation, listing your name first.

Name: Percent Contribution:

_____   _____

_____   _____

_____   _____

Describe your contribution to the completion of the investigation and the writing of the report.

Return this sheet to your lab instructor.

Investigation 3
How Should Data Trends Be Presented?

Introduction

Graphing is a very valuable tool in chemistry. A graph can display a large collection of data at a single glance in a visual format that reveals trends or relationships. Graphs are often used in scientific reports to communicate the results of an investigation in a concise fashion.

Suppose a beverage company wants to compare the cooling behavior of hot beverages in different containers. To keep the work simple, the company has asked you to collect and graph temperature data for the cooling of hot water in two beverage containers. From that data the company wants you to write a report summarizing your work and produce high-quality graphs to accompany the report. If your report and graphs meet the standards of the company, more contracted work could come your way.

Goals

As you complete this investigation you will:

1. Collect temperature data to compare the cooling of hot water in different insulated containers.
2. Transfer the data to a graphing software package (and/or produce a graph of the data by hand as directed by your instructor).
3. Manipulate the data to create two or more professional-quality graphs.

Materials

Source of hot water
Various containers for hot beverages
Thermometer (or temperature probe)
Graphing software
Graph paper (if needed)

Getting Started

Design an experiment to collect temperature and time data for the cooling of a sample of hot water. You should select appropriate time intervals and determine when you will stop collecting temperature data. You should plan at least two trials, preferably three, for each container.

Make sure you know how to use the tools available to you to produce a professional-quality graph. If you are using a computer to produce your graphs, then you should know how to:

1. manipulate entire columns of data with one formula.
2. change the scale of either axis.
3. edit the axis labels and the title.
4. size the graph to fit an entire page.
5. print the graph.
6. do other interesting manipulations suggested by your instructor.

Enter the data into the graphing software available to you. (If your data is already stored on a computer or calculator, then automatic transfer should be possible. See Appendix A for help with data transfer.) Prepare a graph of the temperature (y-axis) versus the time (x-axis), in seconds. Be sure to label your axes and give your graph a title. See Appendix G for other guidelines about producing a professional-quality graph. Also see Appendix G for instructions on preparing a graph by hand (if this is necessary, or required by your instructor).

Since the beverage company didn't specify a temperature scale, prepare a second graph using a different temperature scale than before. If you used Celsius or Kelvin in your first graph, then use the corresponding Fahrenheit values in your second graph. If you used Fahrenheit, then use Celsius for the second graph. Note that you do not have to calculate each individual temperature value if you are using a computer to produce your graphs. You can assign a formula to a vacant data column to create a new set of data. The formula will convert all values of temperature and display them in the new column. (See Appendix G for instructions about using spreadsheet programs or consult your instructor.)

Repeat your procedure using another container. With the instructor's permission, collect data from other groups' experiment. Compare the cooling behavior of the water among the containers.

Report

Submit a report as directed by your instructor. In your report, discuss and compare the cooling behavior of water in the two containers you tested. Also, compare your results to those obtained by other groups in the lab. Pay special attention to the quality of your graphs. The report will be returned if it is unacceptable.

Caution:
While working in the laboratory wear your goggles at all times.
To avoid possible burns be careful handling the hot water.

How Should Data Trends Be Presented? **Team Contribution Form**

Identify the percent contribution each team member made in this investigation, listing your name first.

Name: Percent Contribution:

_____   _____

_____   _____

_____   _____

_____   _____

Describe your contribution to the completion of the investigation and the writing of the report.

Return this sheet to your lab instructor.

Investigation 4
How Is Lab Equipment Used?

Introduction

When planning a given experiment, equipment and techniques used to collect data should offer approximately the same precision. The overall precision of an experiment is determined by the least precise of the measurements you make. Therefore, if you use a piece of equipment that is known to be imprecise, you should not waste time by making other measurements more precisely.

Suppose the supplier of the equipment in your instructional laboratory has recently contracted for your team's services to test the volumetric precision and accuracy of items they have supplied. The equipment includes burets, pipets, volumetric flasks, graduated cylinders, graduated beakers, medicine droppers, and other equipment as directed by your instructor. The manufacturing equipment for each of these lab items is of unknown quality so the measurement ability of these devices is suspect. Assuming that the balance in your lab has been recently calibrated, it is the only piece of equipment you can trust.

Goals

As you complete this investigation you will:
1. Determine proper procedures to weigh equipment and chemicals.
2. Determine proper procedures to measure out specific volumes of liquids.
3. Determine the measurement precision and accuracy of the items sent by the customer.
4. Compare the number of significant figures in values measured with different techniques.

Materials

Water	Droppers
Electronic balance	Sand
Weighing bottles, papers, and/or boats	1-L beaker
Beakers, volumetric flasks, graduated cylinders, pipets, burets, and other volumetric glassware	Other supplies by request

Getting Started

In this investigation your group has to develop a procedure to determine the precision and accuracy of various volume measuring equipment. Because the only item in the lab that you can trust is a balance, your procedure will have to involve a relationship between mass and volume. (Hint: The density of water as it varies with temperature can be found in Appendix J.) As you develop your procedure, keep in mind that you must test both the precision and accuracy of the volume measuring equipment as directed by your instructor. How many trials should you run for

Investigation 4

each item? To become adept at using the available balances you should develop a procedure for weighing out exactly 1 gram of sand. If you have more than one type of balance in your lab, learn how to use all that are available. (Treat the sand as though it were a corrosive chemical, such as sodium hydroxide.) *Chemicals should never be weighed directly on a balance pan!*

Once you have studied the accuracy and precision of the volume measuring items, develop procedures for measuring exact quantities of liquids. Specifically, how would determine the volume and mass of 1 drop of water? How would you measure exactly 10 mL of water using different lab equipment. How would you measure exactly 100 mL of water? Try different procedures to determine what works best. (Treat the water as though it were a corrosive chemical, such as sulfuric acid.)

Other Activities

To further explore uses of lab equipment, complete the tasks below as directed by your instructor. Comment on these items in an appendix to your lab report.

1. Develop a procedure for weighing exactly 1 liter of water in a large beaker. What factors cause difficulties in carrying out this procedure with the equipment available in the lab?
2. Suppose a hospital supply company that prepares solutions for use in intravenous (IV) bags has asked for your help in the preparation of saline (sodium chloride) solutions. A typical IV bag contains 0.900 mass percent sodium chloride. The company usually begins from a solution that is 9.00 mass percent sodium chloride. It wants your help in developing a procedure for preparing 10.0 kilograms of 0.900% solution from the 9.00% starting solution. How would you tell them to prepare the saline solution? Your procedure should include recommendations about the proper measuring equipment.

Report

The report must be typed and grammatically correct. Within the appropriate sections of your report, be sure to address all the questions asked in this investigation. Your report may be returned to you for corrections if it is not acceptable.

Reports should be constructed according to the format described on pages 4 and 5 in this manual, to the extent that this format is appropriate to the experiments that you designed. The emphasis of this report is on your recommendations regarding laboratory procedures so the *Experimental* section should be clear. When writing the report keep in mind the goals of the investigation — in this case, recommendations to a supply company regarding use of its equipment. You should include an explanation about the circumstances under which each item should be used. Finally, you should compare your results to those obtained by other students.

Caution:
While working in the laboratory wear your goggles at all times.

Investigation 4

How Is Lab Equipment Used? **Team Contribution Form**

Identify the percent contribution each team member made in this investigation, listing your name first.

Name: Percent Contribution:

_____   _____

_____   _____

_____   _____

_____   _____

Describe your contribution to the completion of the investigation and the writing of the report.

Return this sheet to your instructor.

Investigation 5
What's in the Bottles?

Introduction

Among the basic problems which frequently confront a chemist is the analysis of materials to determine both the identity of the elements or groups of elements present in a given substance (*qualitative analysis*) and how much of each is present (*quantitative analysis*). In this investigation, we will focus only on the identification of substances. Chemists often must depend on the observation of the chemical properties of substances in order to make an identification, just as a detective often must depend on the observation of the behavior of individuals suspected of being involved in a crime. The chemist may look for clues concerning the identity of a substance by observations of pH, color, odor, viscosity, and the results of mixing two substances. Typical results might be evolution or absorption of heat, formation of an insoluble solid (*precipitate*), color of the precipitate, evolution of a gas, odor of the gas, and so on.

Suppose your team of chemical investigators is working for an environmental agency. You have been asked to identify the contents of several unlabeled bottles collected in the laboratory of a company that is not complying with standards of correctly labeling chemical supplies. The environmental agency wants your team to determine the contents of each bottle obtained from the laboratory and report your findings.

Goals

As you complete this investigation you will:
1. Observe chemical reactions arising from the mixing of solutions of different chemicals.
2. Use your observations to identify the contents of unlabeled bottles, based on the type of chemical reactions expected.
3. Recognize various classes of chemical reactions.
4. Report your findings in a scientific manner.

Materials

Set 1: $AgNO_3$(aq), $Mn(NO_3)_2$(aq), $Ba(NO_3)_2$(aq), HCl(aq), NaOH(aq)
Set 2: $Zn(NO_3)_2$(aq), $Al(NO_3)_3$(aq), $AgNO_3$(aq), NaOH(aq), NH_3(aq)
Set 3: $AgNO_3$(aq), $Ba(NO_3)_2$(aq), HCl(aq), H_2SO_4(aq), NaOH(aq)
Set 4: $AgNO_3$(aq), $Pb(NO_3)_2$(aq), HCl(aq), NH_3(aq), H_2O(aq)
Litmus paper, universal indicator, or pH meter
Spot-plate
Droppers
Thermometer or temperature probe
Other supplies by request

Investigation 5

Getting Started

The purpose of the experiment is to identify the contents of each bottle, matching the bottle numbers or letters with the chemical species listed for your set. You will be assigned one of the four sets of unlabeled bottles previously listed. The sets might be given a unique letter, number, or color code. The bottles will be identified only by code numbers or letters assigned by the environmental firm. Your first task might be to identify the particular set you have been assigned. Because the environmental firm has limited resources, the only reagents you have available are the contents of the unlabeled bottles. However, you can make observations of the interactions among the contents of the bottles by mixing them with one another in drop-by-drop quantities. From an analysis of the observed results of interaction between the bottle contents, you should be able to identify each solution. If you feel that you need some external reagent to confirm your conclusions, consult your instructor and obtain permission first.

Keep in mind that all the unknown substances listed above dissociate in water. For example, when solid silver nitrate, $AgNO_3$, dissolves in water it produces Ag^+ and NO_3^- ions. However, some substances do not dissociate in water. You might consult a table of solubility rules in your text to determine which combinations of substances will give a solid precipitate.

You might consider using litmus paper to determine if your solutions are acids or bases. Defined simply, an acid is a substance that donates H^+ ions to solution when the substance is in water. A base is a substance that accepts H^+ ions. Chemists use pH as a measure of the acidity or basicity of a substance. A neutral solution has a pH of 7, an acid has a pH below 7, and a base has a pH above 7. You can use red and blue litmus paper to provide a simple qualitative way of determining if a solution is an acid or a base. Red litmus paper will turn blue in the presence of base and blue litmus paper will turn red in the presence of acid. If needed, consult your text for further information regarding acids and bases.

Report

The report must be typed and grammatically correct. It may be returned to you for corrections if it is not acceptable. Keep in mind the context for which you are completing this investigation. A discussion of the chemical reactions you observed and equations that represent those reactions must be embedded in an appropriate section of your report. You should also identify each chemical reaction according to its reaction class (single displacement, double displacement, and so on). Also, write net ionic equations and identify spectator ions when appropriate.

> **Caution:**
> **While working in the laboratory wear your goggles at all times.**
> **You are working with strong acids and bases that can cause**
> **permanent damage to eyes and skin.**

What's in the Bottles? Team Contribution Form

Identify the percent contribution each team member made in this investigation, listing your name first.

Name: Percent Contribution:

_____   _____

_____   _____

_____   _____

_____   _____

Describe your contribution to the completion of the investigation and the writing of the report.

Return this sheet to your lab instructor.

Investigation 6
How Can the Waste Be Made Useful?

Introduction

Many chemical processes involve the production of byproducts that may not have immediate use. Because of the impact these substances may have on the environment in their disposal, chemists try to reduce their production, use processes that produce environmentally innocuous substances, or turn them into products that have practical use. In this investigation your group will work with some solutions of substances that when combined may have some practical use.

Suppose your team of savvy investigators has been hired by a chemical waste disposal company. For a substantial fee, the company collects large volumes of waste solutions. As much as possible, waste treatment companies try to concentrate the solutions. In addition, solid waste is easier to dispose of than solutions. As an alternative to disposal, the company thinks that it can generate more revenue by producing something useful from its acquired solutions. Your job is to design procedures for making useful solids from a list of solutions. These solids can then be sold by the company for additional profit.

Goals

As you complete this investigation, you will:
1. Become acquainted with the solubility rules for ionic compounds.
2. Use the solubility rules to design syntheses of several ionic compounds.
3. Separate the solid product from the solution.

Materials

0.1 M solutions of $AgNO_3$, $Mn(NO_3)_2$, $BaCl_2$, HCl, NaOH, H_2SO_4, $Zn(NO_3)_2$, $Al(NO_3)_3$, and Na_2CO_3
Test tubes
Centrifuge and/or filtration equipment
Well plates (if available)
Droppers
Other supplies by request

Getting Started

Your instructor will assign the solids your group will study. You will design experiments for the synthesis and separation of your assigned solids. These solids must be prepared from the solutions supplied by the chemical waste company. On completion of your syntheses you should present the pure solids to your instructor for inspection. The presence of other ions would be considered a contamination of the solid.

Investigation 6

All of the acquired waste solutions consist of dissolved ions. You may want to consult the solubility rules in Appendix J to determine whether your assigned solids are water soluble. Information about filtration and centrifugation in Appendix G may be of some use. For each solid determine how much product is obtained.

Report

The report must be typed and grammatically correct. It may be returned to you for corrections if it is not acceptable. Your report should include thorough discussions of your syntheses, complete with chemical equations. Molecular, complete ionic, and net ionic equations may be useful in describing the chemistry involved in your syntheses. In addition, your report should include recommendations for possible uses of the solids your group synthesized so the chemical waste company can identify potential customers.

> **Caution:**
> **While working in the laboratory wear your goggles at all times.**
> **You are working with strong acids and bases that can cause**
> **permanent damage to eyes and skin.**

Investigation 6

How Can the Waste Be Made Useful? **Team Contribution Form**

Identify the percent contribution each team member made in this investigation, listing your name first.

Name: Percent Contribution:

_____   _____

_____   _____

_____   _____

_____   _____

Describe your contribution to the completion of the investigation and the writing of the report.

Return this sheet to your lab instructor.

Investigation 7
Is the Water Hard or Soft?

Introduction

Homeowners frequently hear from telemarketers peddling everything from magazines to vacuum cleaners. Telemarketers for water purification systems seem especially common lately. They offer free testing of household water in hopes that they can show that an expensive water softener should be purchased. A consumer advocate wishing to help homeowners test their own water has asked for your group's help in developing a technique for testing water hardness at home.

To determine if a home really needs an expensive new water softener, you should address several questions. How do you know whether your water is hard or soft? How hard is it? Which chemicals work to soften the water? As a group, your team will investigate the above concerns. The technique you develop might be endorsed by a consumer advocate company.

Goals

As you complete this investigation you will:
1. Develop a method for determining the hardness of a water sample.
2. Use this method to measure the temporary and permanent hardness of the water.
3. Use this method to measure the relative effectiveness of various chemicals as water softeners.

Materials

0.00100 M $CaCl_2$ solution, which has a hardness of 100 ppm
Soap solution
Water sample
Distilled water sample
Borax, $Na_2B_4O_7(s)$
Sodium chloride, $NaCl(s)$
Sodium carbonate (washing soda), $Na_2CO_3(s)$
Sodium phosphate, $Na_3PO_4(s)$
Sodium sulfate, $Na_2SO_4(s)$
Plastic droppers
Balances
Graduated cylinders, burets, pipets, or other volume measuring devices
Other supplies by request

Investigation 7

Getting Started

Water that contains the impurities Ca^{2+}, Mg^{2+}, and sometimes Fe^{3+}, along with the anions Cl^-, SO_4^{2-}, and HCO_3^-, is called *hard water* because the metal ions cause an undesirable precipitate of soap scum when soap is added. For example, some soaps contain compounds of the anion, $C_{17}H_{35}CO_2^-$. This anion can react with calcium ions present in hard water to make an insoluble compound:

$$2C_{17}H_{35}CO_2^-(aq) + Ca^{2+}(aq) \rightarrow Ca(C_{17}H_{35}CO_2)_2(s)$$

Lathering will not occur until the metal ions associated with water hardness are consumed. The formation of a permanent lather can be used as a measure of the hardness of the water. When developing your method for determining hardness, however, be sure to consider whether suds would form with the first addition of soap in the absence of any hardness. You might begin by comparing the hardness of tap water and deionized or distilled water. A solution of 0.00100 *M* calcium chloride is provided. This solution has a hardness of 100 part per million. This solution can be used as a hard water standard to quantify your results.

Hardness is considered to be permanent if it remains after water is boiled. Hardness that is removed by boiling is called *temporary hardness*. The temporary hardness is caused by bicarbonate compounds, which are converted to insoluble carbonate compounds by heating:

$$Ca^{2+}(aq) + 2HCO_3^-(aq) + heat \rightarrow CaCO_3(s) + CO_2(g) + H_2O(l)$$

Since the calcium is removed as a carbonate precipitate, it is no longer present as free calcium ions in solution. Permanent hardness is caused by chloride or sulfate salts, which are not changed by heating.

When testing the use of various solid chemicals as water softeners, be sure to compare equal amounts of the solids and make sure that the solids are completely dissolved.

Report

Submit a report in the required scientific fashion to your instructor. It may be returned to you for corrections if it is not acceptable. In addition to the usual items, your discussion should address the different types of hardness and the relative efficiency of the various salts as water softeners. You should also compare various water-softening systems, such as the new magnetic systems and older methods such as ion exchange, distillation, and reverse osmosis. Could you use inexpensive chemicals to soften the water?

> **Caution:**
> **While working in the laboratory wear your goggles at all times.**

Is the Water Hard or Soft? **Team Contribution Report Form**

Identify the percent contribution each team member made in this investigation, listing your name first.

Name: Percent Contribution:

_____   _____

_____   _____

_____   _____

_____   _____

Describe your contribution to the completion of this investigation and the writing of the report.

Return this sheet to your lab instructor.

Investigation 8
How Hot Is the Water?

Introduction

All physical and chemical processes are accompanied by changes in energy, often in the form of heat. For example, when a substance cools it loses heat to the environment. Changes in energy are guided by the *first law of thermodynamics*, which states that the energy of the universe is constant. You will use this law to solve the problem posed in this investigation.

Suppose you have recently been put on a technical service team to help your company's customers with your product line. One customer recently returned the coffee pot you manufacture, claiming that the temperature of the water is not hot enough to brew coffee. Your team must measure the temperature of the water inside the pot. Unfortunately, the only thermometer you have available to you has no marks above 40°C. Using only this thermometer and other materials in your lab (except another thermometer), devise and carry out methods to determine the temperature of the water inside the pot.

Goals

As you complete this investigation you will:
1. Apply the principles of thermochemistry to solve your company's problem.
2. Design two alternative approaches to solving the problem.
3. Report your results in a scientific manner.

Materials

Coffeepot
40°C thermometer
Styrofoam cups
Stirrers
Graduated cylinders, burets, pipets, or other volume measuring devices
Balances
Ice
Cold water
Samples of various metals
Other supplies by request

Investigation 8

Getting Started

You may need to consult your textbook about heat transfer, specific heat, and the first law of thermodynamics. Design an experiment that will allow you to determine the temperature of the water in the pot with the 40°C thermometer. You may do anything you wish to bring the water temperature into the range of the thermometer, provided that the measurements you take can be used to calculate the original temperature of the coffeepot water. When you collect your hot water samples, make sure the coffee pot has been on long enough for them to be at its highest temperature. Consider precautions that you might take to prevent the water from cooling off before you start your procedure. Run several trials to confirm the validity of your technique and the data you obtain. Devise a second method for determining the temperature of the water.

Report

Submit a report written in the usual scientific fashion. It may be returned to you for corrections if it is not acceptable. In addition to describing your methods and results, describe your group's alternative approach to solving the problem and the data you expect to obtain from that approach.

> **Caution:**
> **While working in the laboratory wear your goggles at all times.**
> **Be careful handling the hot water to avoid burns.**

Investigation 8

How Hot Is the Water? Team Contribution Report Form

Identify the percent contribution each team member made in this investigation, listing your name first.

Name: Percent Contribution:

_____   _____

_____   _____

_____   _____

_____   _____

Describe your contribution to the completion of this investigation and the writing of the report.

Return this sheet to your lab instructor.

Investigation 9
Which Metal Will Burn the Skin?

Introduction

All physical and chemical processes are accompanied by changes in energy, often in the form of heat. For example, when a substance cools it loses heat to the environment. Changes in energy are guided by the *first law of thermodynamics*, which states that the energy of the universe is constant. You will use this law to solve the problem posed in this investigation.

You have recently been put on a parks and recreation committee that has been placed in charge of selecting new park benches. Although park bench design involves consideration of several factors for materials selection, the mayor wants your team to study benches that can be placed in sunny areas but won't burn the skin. To address the mayor's restrictions your group will have to study heat transfer for various metals used in the construction of park benches.

Goals

As you complete this investigation you will:
1. Apply the principles of thermochemistry to study heat transfer for various metals.
2. Make comparisons between the microscopic structure of the metals and their observed specific heats.
3. Report your results and recommendations to the contractor.

Materials

Thermometer or temperature probe
Styrofoam cups
Stirrers
Graduated cylinders, burets, pipets, or other volume measuring devices
Balances
Ice
Cold water
Samples of various metals
Other supplies by request

Getting Started

You may need to consult your textbook about heat transfer, specific heat, and the first law of thermodynamics. In general, you can use first law relationships to determine the specific heat of one substance from information known about another. You might begin by designing an

Investigation 9

experiment that involves heating your metal to a certain temperature, then adding it to another substance at a lower temperature. (Alternatively, you could cool the metal, then add it to another substance at a higher temperature.) Consider precautions that you might take to prevent the metal from cooling off (or heating up) before you add it to the other substance. Devise a method for measuring heat transfer and specific heat of as many metals as possible using available temperature measuring devices. You should run multiple trials to confirm the reliability of your data. Determine relationships, if any, among the quantity of metal, its microscopic structure, and the measured specific heat. If your data is inconclusive you might consider studying other properties of the metal, such as density. The metal samples sent by various park bench manufacturers might not be adequately labeled so you may need to determine the composition of them from your data. Consult your textbook or the Internet for needed information.

Report

Submit a report in the required scientific fashion to your instructor. In addition to the usual items, your discussion should address relationships you identified between the composition of the metals and their observed specific heats. You should also compare the actual specific heats to your experimental values. Based on your results, make recommendations to the mayor about the park bench metals that are most appropriate for preventing burns to the skin. Describe other thermal factors besides specific heat that should be considered in selecting a metal for park benches.

Caution:
While working in the laboratory wear your goggles at all times.
Be careful handling the hot objects to avoid burns.

Investigation 9

Which Metal Will Burn the Skin? Team Contribution Report Form

Identify the percent contribution each team member made in this investigation, listing your name first.

Name: Percent Contribution:

_____   _____

_____   _____

_____   _____

Describe your contribution to the completion of this investigation and the writing of the report.

Return this sheet to your lab instructor.

Investigation 10
Are All Neutralization Reactions the Same?

Introduction

All chemical reactions are accompanied by changes in energy, often in the form of heat. For example, when a neutralization reaction occurs energy is released to the environment. Changes in energy are guided by the *first law of thermodynamics*, which states that the energy of the universe is constant. You will use this law to solve the problem posed in this investigation.

An environmental firm has contracted for your team to investigate a phenomenon it has observed in neutralizing acid solutions. The firm has found that heat is released in the neutralization of acids that is contributing to thermal pollution in the areas surrounding its facility. It does not have the equipment or the intellectual know-how to understand what is happening so it is asking your group to study the neutralization reactions for various acids at multiple concentrations. Specifically, the environmental firm wants to know if the heat released during the neutralization varies with different acids.

Goals

As you complete this investigation you will:
1. Apply the principles of thermochemistry to determine the heats of neutralization for various acids.
2. Make comparisons of the observed heats of reaction for the acids analyzed.
3. Report your results and recommendations to the environmental firm.

Materials

Thermometer or temperature probe
Styrofoam cups
Stirrers
Balances
Ice
Cold water
HCl, HNO_3, H_2SO_4, and $HC_2H_3O_2$ at various concentrations[*]
2 M NaOH[*]
Graduated cylinders, burets, pipets, or other volume measuring devices
Other supplies by request
    [*]*Obtain exact concentration from reagent bottle.*

Getting Started

You may need to consult your textbook about neutralization reactions and the first law of thermodynamics. You might begin by designing an experiment that involves determining the heat released from the reaction of the acids with a neutralizing amount of base. Devise a method for determining the heat released upon neutralization for as many acids and concentrations as possible using available temperature measuring devices. You should run multiple trials to confirm the reliability of your data. Determine relationships, if any, among the quantity of acid, the acid's chemical nature, and the amount of heat released. There are several ways of expressing your data for comparative purposes. You could express the heat released on a per–mole–of–acid–reacted basis. Alternatively, the heat released could be expressed in terms of available acidic hydrogens on the acid. Finally, it could be expressed on a per–mole–of–water–formed basis. Compare the values expressed in all these ways for the acids you tested.

Report

Submit a report in the required scientific fashion to your instructor. In addition to the usual items, your discussion should address relationships you identified between the concentration of the acid and the observed heat of neutralization. In addition, you should compare the heats of neutralization for the different acids you tested and propose explanations for similarities and differences.

Caution:
While working in the laboratory wear your goggles at all times.
You are working with strong acids and bases that can cause
damage to skin and eyes.

Are All Neutralization Reactions the Same? **Team Contribution Report Form**

Identify the percent contribution each team member made in this investigation, listing your name first.

Name: Percent Contribution:

_____   _____

_____   _____

_____   _____

_____   _____

Describe your contribution to the completion of this experiment and the writing of the report.

Return this sheet to your lab instructor.

Investigation 11
How Much Sodium Bicarbonate Is in the Mixture?

Introduction

When chemists use materials in various processes, such as the synthesis of new substances, it is convenient to use pure materials. However, the purification of materials, whether by physical or chemical methods, is often quite expensive. Sometimes it is much cheaper to use materials that contain impurities; that is, the material might be a mixture of two or more substances. If substances with impurities are to be used in a chemical process, however, it is necessary to know how much of each substance is present in the mixture. Determining the amount of each substance in a mixture is often carried out by chemical means.

A chemical manufacturing company that uses sodium bicarbonate as a precursor to other sodium compounds recently contracted for your team's services to determine the amount of sodium bicarbonate in a mixture known to contain sodium chloride. It wants to know if this is a reasonable mixture for sodium carbonate production. If the mixture contains more than 90% sodium bicarbonate, then the reduced purity of their product is worth the savings on raw material. Using the materials available in your facility you are to devise and carry out methods for determining the relative amounts of each substance in the mixture and report your findings to the manufacturing company.

Goals

As you complete this investigation you will:
1. Devise two or more methods to determine the percent composition of a mixture.
2. Carry out multiple trials of one such method and determine the amounts of $NaHCO_3$ and NaCl in a mixture containing only these two substances.
3. Report your findings to the chemical manufacturing company.

Materials

Sodium chloride, pure
Sodium bicarbonate, pure
Unknown mixture of sodium chloride and sodium bicarbonate
1 M hydrochloric acid
1 M sulfuric acid
1 M sodium hydroxide
Balances and other common laboratory equipment
Other supplies by request

Investigation 11

Getting Started

You will be given a white powder, which is a mixture of two solids: sodium bicarbonate (NaHCO$_3$) and sodium chloride (NaCl). You are to determine the percentages by mass of each solid in the mixture. Devise at least two chemical procedures for determining the composition of the mixture. Select the one you think has the greatest chance of success and try it out on the pure substances, and on a mixture that you make from the pure substances. Using the reagents provided, you might begin by studying the chemical properties of the pure substances. Note that the solutions provided are not standardized. This may impact the procedure your group devises. As you design your experiments, how will you know when the reactions come to completion? If you decide to add reagents to the mixture, how will you know how much to add? You could also consult your textbook or other resources about the chemical behavior of the mixture components. Apply your method to the unknown. If time permits, try the alternative method your group designed.

Report

Following the guidelines for writing a complete report, describe your method and the results obtained. Compare your results to those other teams obtained. Be sure to report the results on the percent composition to a number of significant figures that is suitable for the data that you collected. Also, describe an alternative method for determining the mixture composition. Based on your results, what recommendation will you make to the chemical manufacturing company? If directed to do so by your instructor, you might consult information on the Internet to compare prices of sodium bicarbonate at various purities.

> **Caution:**
> **While working in the laboratory wear your goggles at all times.**
> **You are working with strong acids and bases that can cause**
> **damage to skin and eyes.**

How Much Sodium Bicarbonate Is in the Mixture? **Team Contribution Form**

Identify the percent contribution each team member made in this investigation, listing your name first.

Name: Percent Contribution:

_____   _____

_____   _____

_____   _____

_____   _____

Describe your contribution to the completion of the investigation and the writing of the report.

Return this sheet to your lab instructor.

Investigation 12
Is It Economical to Recycle Aluminum?

Introduction

Recycling of discarded materials can generally take two forms. First, the old materials can simply be reused. A second way of recycling is to convert the used material into another useful product. For example, through a series of chemical reactions aluminum cans can be converted to a compound known as alum which has many uses. Rather than selling its materials to other manufacturers, an aluminum recycling company wants to produce useful alum products of its own. Specifically, it believes that it can break into the personal care products business by devising a cheap way of manufacturing alum, $KAl(SO_4)_2 \cdot 12H_2O$, an active ingredient in deodorants. The company has contracted for your team to investigate the chemistry associated with the preparation of alum from aluminum cans. It wants you to devise a cost-effective route to the synthesis of alum with an acceptable product yield.

Goals

As you complete this investigation you will:
1. Devise and optimize a method for the synthesis of alum.
2. Determine the percent yield and cost effectiveness of the method.
3. Report your findings and recommendations to the aluminum recycling company in an appropriate manner.

Materials

Aluminum foil or cans
KOH pellets
9 M H_2SO_4
Ethanol
Filter paper

Scissors
Sandpaper
Büchner funnel
Other supplies by request

Getting Started

As you plan your experiment consider the reagent quantities you will need to adequately understand the reaction. A slight excess of reagent may be required to completely dissolve the aluminum in a timely manner. If you will use aluminum cans, keep in mind that they have an outer coating of paint and an inner plastic coating. You may have to take precautions to prevent introduction of these coatings into the reaction mixture. If you do not, you may have to remove solid debris from the reactant mixture. The recycling company wants you to conduct your experiments on a small scale (approximately 1.0 to 1.5 g samples of aluminum). It does not wish to absorb any costs associated with disposing of excess reagents.

Investigation 12

The preparation of alum proceeds through a series of steps that begins with the reaction of aluminum and potassium hydroxide. Gently heating the mixture will greatly speed up this first step:

$$2Al(s) + 2KOH(aq) + 6H_2O(l) \rightarrow 2K^+(aq) + 2Al(OH)_4^-(aq) + 3H_2(g) \quad (1)$$

Caution: Potassium hydroxide readily dissolves human tissue.

Addition of sulfuric acid to the cooled product mixture of reaction (1) is the second step of the series:

$$2Al(OH)_4^-(aq) + H_2SO_4(aq) \rightarrow 2Al(OH)_3(s) + SO_4^{2-}(aq) + 2H_2O(l) \quad (2)$$

Caution: 9 *M* sulfuric acid is a strong, concentrated acid.

Addition of excess sulfuric acid dissolves the aluminum hydroxide:

$$2Al(OH)_3(s) + 3H_2SO_4(aq) \rightarrow 2Al^{3+}(aq) + 3SO_4^{2-}(aq) + 6H_2O(l) \quad (3)$$

Upon cooling the product mixture from (3) in an ice bath, alum will form according to the equation:

$$Al^{3+}(aq) + K^+(aq) + 2SO_4^{2-}(aq) + 12H_2O(l) \rightarrow KAl(SO_4)_2 \cdot 12H_2O(s) \quad (4)$$

Stirring can prevent crystal formation. The product should form nice, readily observable crystals. You may have to scratch the inside of the glass to initiate the crystallization process. (If too much excess sulfuric acid was added, crystals of alum may not form.) Your alum product can be isolated and dried using a suction filtration apparatus. It is sometimes desirable to wash the solid products of a chemical reaction of remaining unreacted chemicals. However, you must be careful not to wash the solid with a solvent that might dissolve your product. You might try a 1:1 mixture of ethanol and water to wash your alum crystals. After your product is dry, calculate your percent yield. You should perform the complete synthesis of alum from aluminum at least two times to refine your technique. If time permits, a third synthesis is preferable.

Report

Write a report in the usual scientific manner, describing your method and the results obtained. As part of this investigation your group must complete a thorough cost analysis for production of alum from aluminum cans or foil. These costs should include raw materials, chemical reagents, filtering supplies, and waste disposal. In your analysis, don't forget to include labor costs and the costs of the preliminary work carried out by your group. What happens to the costs when the process is scaled up to the production of 100 kg of alum? You may have to do an Internet search to complete the cost analysis part of this investigation. Be sure to cite your sources.

> **Caution:**
> While working in the laboratory wear your goggles at all times.
> You are working with concentrated acids and bases that can
> cause severe damage to tissue, skin, and eyes.

Is It Economical to Recycle Aluminum? **Team Contribution Form**

Identify the percent contribution each team member made in this investigation, listing your name first.

Name: Percent Contribution:

_____   _____

_____   _____

_____   _____

_____   _____

Describe your contribution to the completion of the investigation and the writing of the report.

Return this sheet to your lab instructor.

Investigation 13
What Is a Copper Cycle?

Introduction

A company that mines platinum has recently discovered a rich source of copper near one of its mines. The company has asked your savvy chemical team to probe the chemistry of copper and some of its compounds. By understanding the chemistry of copper it hopes that it can easily separate the copper from platinum and other precious metals it mines. Specifically, the company thinks that by starting with copper metal and making copper(II) hydroxide, copper(II) nitrate, and copper(II) sulfate (not necessarily in that order) in some sequence of steps, copper metal can be recovered. If your cyclic synthetic route is successful, the company believes that it can separate the precious metals from the copper during one or more of the steps.

Your team must devise a synthetic route that is cyclic, allowing you to study a series of reactions that ultimately recovers the used copper. The company does not have a chemist on staff so there is not a qualified person available to explain the chemical processes occurring in the production of the copper compounds. The company insists that your report include a careful description of the reaction types that occur as you synthesize the compounds of interest.

Goals

As you complete this investigation you will:
1. Probe the chemical behavior of copper.
2. Synthesize various copper compounds and describe the chemical reactions taking place.
3. Recover the copper metal from the series of reactions that produces the chemicals of interest.
4. Report your findings to the mining company.

Materials

Acetone
Copper wire
3 M HCl(aq)
6 M HNO$_3$(aq)
6 M NaOH(aq)
2 M H$_2$SO$_4$(aq)

Zinc
Büchner funnel
Litmus paper
Filter flask
Other supplies by request

Getting Started

Develop procedures for synthesizing the compounds of interest in a cyclic manner. Solubility rules and a metal activity series might be useful. You should prepare the materials using small quantities of starting material to reduce the amount of waste and to minimize the costs (raw

Investigation 13

material and waste disposal) to the customer. In addition to your syntheses, you must include a route to recovering native copper which can be returned to the company. Upon completion of a proposal you must obtain your instructor's approval before carrying out experimentation.

You should begin by dissolving a piece of copper metal (not to exceed 0.5 g to conserve the metal) in a suitable amount of nitric acid as directed by your instructor. Alternatively, your instructor might ask you to calculate the amount of nitric acid needed from stoichiometric relationships. You may have to add additional reagent to force the reaction to completion. Careful heating will also help dissolve the copper metal. **Be very careful; you are working with strong acids and bases. Also, the reaction of copper metal with nitric acid produces toxic NO_2 gas. This reaction should be run in a hood.** The products of this reaction are copper(II) nitrate, nitrogen dioxide, and water. For all the reactions, add the reagents slowly with continuous stirring between additions. As the reactions proceed, make as many observations about the chemical behavior as possible, identifying all products formed. Consider other comments listed below:

- Some products may not be formed as a solid. How can products in solution be isolated?
- When you prepare copper(II) hydroxide be careful when adding sodium hydroxide. You should add this reagent slowly until the solution just turns basic. If you add the hydroxide too fast you will form black copper(II) oxide instead of copper(II) hydroxide.
- The hydrochloric acid listed previously is available only to remove excess zinc that might be present if you use it as one of your reagents. The HCl reacts with the zinc to form hydrogen gas and zinc chloride. After addition of the HCl you should heat the mixture slightly to make sure all the hydrogen gas has been driven off. **Be very careful** — the hydrogen could ignite if brought too close to the heat source.
- At the end of your experimentation you should have recovered all the copper. You can use a Büchner funnel with a filter flask to dry your copper, then wash it with a little acetone. **Acetone is flammable** so be careful! You should compare your final amount of copper to the amount of starting material by calculating the percent recovered. For each trial compare the percent recovery.

Report

You should write your report in customary scientific fashion. Your report to the customer should include careful analyses of the reactions you ran. For each synthesis describe the reaction type, write chemical equations, and explain your observations from a microscopic standpoint. Where appropriate, you should write molecular, ionic, and net ionic equations.

> **Caution:**
> **While working in the laboratory wear your goggles at all times. You are working with strong acids and bases that can damage skin and eyes. You are preparing gases that may be flammable or toxic, so all reactions should be run in a fume hood.**

What Is a Copper Cycle? **Team Contribution Form**

Identify the percent contribution each team member made in this investigation, listing your name first.

Name: Percent Contribution:

_____   _____

_____   _____

_____   _____

_____   _____

Describe your contribution to the completion of the investigation and the writing of the report.

Return this sheet to your lab instructor.

Investigation 14
Who Wrote the Ransom Note?

Introduction

Criminal investigators use chemistry routinely to solve crimes. One important technique in a forensics lab, *chromatography*, involves the separation of mixtures found in evidence at a crime scene. For example, sometimes a piece of evidence might be an ink sample. Forgery cases, kidnapping cases (involving a ransom note), and other crimes involving writing are often investigated using an analytical tool called chromatography. Chromatography is a general term referring to the separation of a mixture. In this investigation, your team will be given a sample of ink found at the scene of a crime. Your job will be to determine which pen made the ink mark.

Goals

As you complete this investigation, you will:
1. Learn about the technique of paper chromatography.
2. Develop a paper chromatography technique to separate components in an ink sample.
3. Identify an ink sample based on its chromatogram.
4. Write a report that could be used as evidence in a criminal trial.

Materials

Black ink pens	Plastic wrap or beaker cover
Chromatography paper	Ink sample from "crime scene"
Scissors	Deionized water
Large beaker	20%, 50%, and 90% ethanol solutions
Rulers	Other supplies by request

Getting Started

Paper chromatography is one type of chromatography used to separate a mixture of components (such as ink) to observe its component substances. Black ink is usually a mixture of several different colors that can be separated with the right chromatographic conditions. Inks from different pens are commonly composed of different components that make the ink unique. To separate the components in an ink sample, a small dot of ink is placed on a strip of chromatography paper about 2.5 cm from the end. The paper is then stood up in a beaker containing about 1 to 1.5 cm of solvent. Why should the solvent depth remain below the level of the ink dots? The chromatography paper soaks up the solvent slowly and the ink migrates with it towards the top of the paper. The individual components of the ink move at different rates and a

Investigation 14

pattern characteristic of the ink appears on the paper.

Designing chromatography experiments usually involves selection of the proper solvent for the system of interest. In this investigation your team of crime specialists has been provided several solutions of ethanol with water. Your first task should be to determine which solvent mixture is appropriate for separation of ink components in the known pens. In addition to the ethanol solutions, you should try using water as the solvent. Why do you suppose some solvent systems work better than others?

Your group will be given one or two samples of ink taken from a crime scene. Because the sample is taken from a limited supply that will be used in court, do not use the unknown until your team is certain of the solvent system you will use.

In addition to determining the proper solvent, you should develop a way of quantifying your data for your report; that is, you might consider ways of expressing your results in numerical terms that allow for easy comparison of results from one experiment to another.

Report

The report must be typed and grammatically correct. It may be returned to you for corrections if it is not acceptable. Your report should include a discussion of factors that may affect the rate at which the solvent and the ink components move up the chromatography paper. Remember that your report may be used in a court of law. If directed by your instructor, include your chromatograms with your report.

> **Caution:**
> **While working in the laboratory wear your goggles at all times.**
> **You are working with solvents that may be flammable.**

Investigation 14

Who Wrote the Ransom Note? Team Contribution Form

Identify the percent contribution each team member made in this investigation, listing your name first.

Name: Percent Contribution:

_____   _____

_____   _____

_____   _____

_____   _____

Describe your contribution to the completion of the investigation and the writing of the report.

Return this sheet to your lab instructor.

Investigation 15
How Can UV Sensitive Beads Be Used to Test Sunscreens?

Introduction

Ultraviolet light from the harsh rays of the sun has led to development of a variety of skin and eye protection products. During the summer, drug store counters often display UV protection lotions that range from 2 to 35 on the SPF scale. In addition to the skin, eyes must be protected from the damaging UV rays of the sun.

A company that develops UV protective products has begun testing a new experimental protocol. They believe that UV sensitive beads can serve as a surface for testing their skin care products. Unfortunately, they have had little success in developing an experimental procedure that makes use of these beads. Because of your laboratory skills, the company has contracted the services of your group to develop a procedure to use UV sensitive beads for testing the efficacy of sunscreens. In addition, the company wants you to use the beads to test the UV protection of sunglasses.

Goals

As you complete this investigation you will:
1. Develop a method for using UV sensitive beads to test sunscreens.
2. Use your method to test and compare various sunscreens.
3. Test various sunglasses for UV protection. (You may have to bring these from home.)

Materials

UV sensitive beads of various colors
Sunscreens ranging in SPF protection
Soap
Cheese cloth or other fabrics
Plastic bags and/or plastic wrap
UV lamp
Balances
Other supplies by request

Getting Started

In this investigation you will develop a procedure for using UV sensitive beads to test sunscreens. You might begin by observing what happens to the beads when exposed to the sun or when placed near a UV lamp. How long does it take for the beads to change color? Is direct exposure to the sun necessary to change the color of the beads? Do you get better results working in the shade? After you get a feel for how the beads behave in the presence of UV light, determine how you might use them to test sunscreens for UV protection. As you design your procedure, consider what variables to control. What will you use as a control for a basis of comparison?

In addition to testing sunscreens, observe what happens when the UV light is filtered through various kinds of glass. For example, what happens when the beads are placed in a glass bottle or under eye glasses? What happens when UV light is filtered through sun glasses?

Report

The report must be typed and grammatically correct. It may be returned to you for corrections if it is not acceptable. Your report should include a discussion of your method for using UV sensitive beads to test sunscreens and sunglasses. Remember that the company wants your recommendations for an experimental protocol.

> **Caution:**
> **While working in the laboratory wear your goggles at all times.**
> **Protect your eyes from the UV lamp.**

Can UV Sensitive Beads be Used to Test Sunscreens? Team Contribution Report Form

Identify the percent contribution each team member made in this investigation, listing your name first.

Name: Percent Contribution:

_____   _____

_____   _____

_____   _____

_____   _____

Describe your contribution to the completion of this investigation and the writing of the report.

Return this sheet to your lab instructor.

Investigation 16
What Factors Affect the Intensity of Color?

Introduction

A company that makes food coloring needs your team's help. The company has contracted with your group to determine the factors that affect the intensity of color. Specifically the company needs background theoretical information about these factors as they relate to their food coloring products. It has sent three samples of their products for your team to investigate using spectroscopy as your technique for analysis. A variety of factors can affect the measured intensity of the color of a solution. In this investigation you will probe some of these factors.

Goals

As you complete this investigation you will:
1. Determine the factors that affect the intensity of the color of a solution as measured by a spectrophotometer or colorimeter. (There is also a non-spectroscopic technique described in Appendix B.)
2. Develop an equation that relates the amount of light transmitted through a solution to the concentration of the solution.
3. Report your findings in the usual scientific manner.

Materials

Red food coloring solution[*]	Volumetric and other lab equipment
Green food coloring solution[*]	Device to measure light transmittance
Blue food coloring solution[*]	Cuvettes
Distilled water	Tissues (lint free) to clean outside of cuvette

[*]*Assign the food coloring concentrations to 1 in arbitrary units.*

Getting Started

The intensity of light transmitted through a solution can be measured with a device called a *spectrophotometer* or *colorimeter*. (For more information about spectroscopy see Appendix B.) Spectrophotometers and colorimeters are designed so that the intensity of transmitted light is proportional to a voltage that can be measured by a meter. When there is a species present that absorbs light, the measured voltage after passing through the sample is different. In order to determine the appropriate conversion equation between transmittance and voltage, the spectrophotometer or colorimeter must be calibrated at 0% and 100% transmittance. (Also see Appendix B for information about calibrating your instrument. Your instructor may also demonstrate the proper way to calibrate your equipment.) After calibration is complete, you can read the percent transmittance (%T) data on the instrument display. For this investigation, be sure you are reading %T, and not some other property of the detected light intensity.

Investigate various factors that affect the amount of light transmitted through the food coloring solutions. These factors should include the wavelength of the light passing through the solution, the concentration of the solution, and the color of the solution. Your instructor may also ask you to investigate the thickness of the solution through which the light is passing. To study concentration you should use at least seven concentrations of the food coloring solution. For each dye, identify the wavelength that gives you the most quantifiable results. What criteria will you use to make this determination?

Using your measurements of transmittance, determine if there is a linear, mathematical relationship between transmittance (or some function of transmittance) and concentration of the solution. A graphing program will be extremely helpful in studying the various suggested functions of %T, listed below, and concentration. (See Appendix G for information about using spreadsheets to generate graphs.) You might begin by graphing %T versus concentration for each of your experiments to see if there are patterns in the data. Plot other functions of %T to see if you can find one or more straight-line relationship. Some possible functions include:

$$(\%T)^2,\ 1/\%T\ (\text{or}\ 100/\%T),\ (\%T)^{1/2},\ \log(\%T),\ \log(1/\%T),\ 10^{\%T}$$

Once you identify a straight-line relationship, develop an equation in the form of $y = mx + b$, where y is the function of %T, x is concentration, m is the slope of the line, and b is the intercept of the line. Why does finding a straight-line relationship help you to understand the nature of the relationship between transmittance and concentration? How might this relationship be put to use?

Report

The report must be typed and grammatically correct. It may be returned to you for corrections if it is not acceptable. Your report must include a discussion of the factors that affect the intensity of color. In the course of this investigation, you will generate many graphs. As directed by your instructor, these graphs should be included with your report. You should consider consolidating some data in a single graph to help make the points you discuss in your report.

> **Caution:**
> **While working in the laboratory wear your goggles at all times.**

What Factors Affect the Intensity of Color? Team Contribution Form

Identify the percent contribution each team member made in this investigation, listing your name first.

Name: Percent Contribution:

_____   _____

_____   _____

_____   _____

_____   _____

Describe your contribution to the completion of the investigation and the writing of the report.

Return this sheet to your lab instructor.

Investigation 17
How Much Cobalt Is in the Soil?

Introduction

A group of dairy farmers has recently reported to its county cooperative extension service that its animals appear sick. Based on the symptoms, the county suspects that the cows are suffering from a cobalt deficiency. Cobalt(II) is required for production of vitamin B_{12}. Proper animal diet requires the soil to contain 0.13 to 0.30 mg cobalt per kg of soil. Soils that do not meet this condition can be amended with careful addition of cobalt nitrate or other cobalt compounds. Because it lacks the proper equipment, the county extension service has contracted for your team to determine the amount of cobalt(II) in a solution sent to your lab. To get representative samples, the extension service collected 2.00 kg of soil from 20 different farms in the county. The cobalt was then extracted from the soil, dissolved in water, and concentrated by a factor of 10,000 to prepare 1 L of solution. Because the cobalt ion is a colored species in water, spectroscopic techniques can be used to determine the concentration of cobalt(II) in an unknown solution.

Goals

As you complete this investigation you will:
1. Prepare a standard plot of absorbance versus known cobalt(II) nitrate concentrations.
2. Determine the concentration of cobalt(II) in the solution obtained from a soil sample.
3. Determine if cobalt nitrate should be added to the soil. If so, how much will be required to meet the necessary nutritional needs of the animals?
4. Report your findings to the cooperative extension service.

Materials

0.1 M^* $Co(NO_3)_2$(aq)
Unknown cobalt(II) solution
Volume measuring equipment
Device to measure light absorbance

Cuvettes
Tissues (preferably lint free)
Other supplies by request

Obtain exact concentration from the reagent bottle.

Getting Started

For information about the basics of spectroscopy, see Appendix B of this manual. In Appendix B the quantity, percent transmittance (%T), is described as a relationship between the intensity of light before and after passing through a solution. Another quantity known as absorbance (A) is related to %T by the following equation:

$$A = \log(100/\%T)$$

The absorbance is related to concentration of a light-absorbing species by the Beer–Lambert Law, $A = \varepsilon bC$, where ε is a proportionality constant called the molar absorptivity that is specific for the absorbing species, b is the thickness of the solution in the light path (usually constant for an experiment), and C is the molarity of the absorbing species in the solution. For a given light-absorbing species, a plot of absorbance versus concentration should give a straight line. This straight–line relationship can be used to determine the concentration in an unknown solution of the species tested. For this investigation, a plot of the absorbance versus concentration of various cobalt(II) solutions can be used to determine the cobalt (II) ion concentration in the unknown sample sent by the county extension service.

Use the instructions in Appendix B for calibrating the spectrometer or colorimeter. (If neither of these devices is available you can use the non-instrumental technique described in Appendix B.) You will also have to determine the wavelength at which you will run your experiments. A solution of 0.1 M cobalt(II) nitrate is available for preparation of your known solutions. (Make sure you write down the exact concentration of the known solution in your lab notebook.) Develop a procedure for using the cobalt(II) nitrate solution to make several dilutions (at least seven). You will have to select an appropriate range of concentrations. Once your solutions are prepared, use the spectrometer or colorimeter to collect absorbance data for each concentration. If a graphing program is readily available, construct a plot of your data that you can use to determine the concentration in the unknown. At high concentrations (that vary from one chemical species to another) there is deviation from the Beer–Lambert Law. Use your plotted data as a guide for identifying the point at which deviation occurs. If you do not see any deviation from linearity, your concentrations could be in the correct range.

Determine if the amount of cobalt(II) in the soil meets the nutritional needs of the animals. If not, you should calculate the amount of cobalt(II) nitrate that must be added to the soil to reach a level of 0.13 to 0.30 mg of cobalt(II) per kilogram of soil.

Report

As usual you should write your report in customary scientific fashion. Include your recommendations to the cooperative extension service regarding amending soil on the county farms. Remember to include appropriate graphs with your report. (Further information about cobalt deficiency can be found by doing an Internet search on the term *cobalt deficiency*.)

> **Caution:**
> **While working in the laboratory wear your goggles at all times.**

Investigation 17

How Much Cobalt Is in the Soil? Team Contribution Form

Identify the percent contribution each team member made in this investigation, listing your name first.

Name: Percent Contribution:

_____   _____

_____   _____

_____   _____

_____   _____

Describe your contribution to the completion of the investigation and the writing of the report.

Return this sheet to your lab instructor.

Investigation 18
How Much Copper Is in the Coin?

Introduction

Starting in 1983, some U.S. mints started to change the composition of the one-cent coin. This change was completed in all mints by 1984. The amount of copper in a penny was modified during that period because of a change in the price of copper. The other metal used in pennies is zinc. Since then, copper prices have changed. A copper mining firm hoping to recoup its losses wants your team to determine the amount of copper and zinc in coins older than 1983 and newer than 1984 and compare the relative raw material costs for producing the coins today. You will also develop a method for disposing of all the waste your team generates during the investigation.

Goals

As you complete this investigation you will:
1. Develop a method for determining the concentration of a copper(II) ion solution.
2. Determine the amount of copper and zinc in one-cent coins minted before 1983 and after 1984.
3. Compare the cost of producing these coins using copper and zinc metals as the raw materials.
4. Develop a method of waste disposal for all the chemicals you generate.
5. Report your findings in the usual scientific manner.

Materials

Pennies of various ages	1.0 M Zn(NO$_3$)$_2$
6 M HNO$_3$(aq)	Balances
6 M HCl(aq)	A device for measuring light absorbance
6 M NaOH(aq)	Cuvettes
Cu(NO$_3$)$_2$·3H$_2$O (solid)	Tissues (preferably lint free)
Solid copper	Volume measuring equipment
Solid zinc	Other supplies by request

Getting Started

With careful heating, copper will dissolve in 6 M nitric acid to produce a blue solution and a reddish-brown gas, according to the following equation:

$$Cu(s) + 4HNO_3(aq) \rightarrow Cu(NO_3)_2(aq) + 2NO_2(g) + 2H_2O(l)$$

The gas is toxic, so this reaction must be carried out carefully in a hood. Caution: Nitric acid is corrosive to the skin. Zinc also reacts with nitric acid which can be described by a similar equation. The reaction can become very slow if there is not enough acid present. Consider what might be a reasonable excess of acid to make the reaction proceed faster.

For information about the basics of spectroscopy, see Appendix B of this manual. In Appendix B the quantity percent transmittance (%T) is described. Another quantity known as absorbance (A) is related to %T by the following equation:

$$A = \log(100/\%T)$$

Absorbance is related to the concentration of a light-absorbing species by the Beer–Lambert Law, **A** = ε**bC**, where ε is a proportionality constant called the molar absorptivity that is specific for the absorbing species, **b** is the thickness of the solution in the light path (usually constant for an experiment), and **C** is the molarity of the absorbing species. For a given light-absorbing species, a plot of absorbance versus concentration should give a straight line. This straight line relationship can be used to determine the concentration of an unknown solution of that species.

Use the instructions in Appendix B for calibrating the spectrometer or colorimeter. (If these are unavailable you can use the non-instrumental technique described in Appendix B.) You will have to determine the wavelength at which to run your experiments. Solid copper(II) nitrate is available for preparation of your known solutions. Consider the other materials listed as you design your experiments. You will have to determine the range of concentrations and develop procedures for using the copper(II) nitrate to make those solutions (at least seven). Collect absorbance data for each concentration you prepare. Construct a plot of your data that you can use to determine the concentration in an unknown solution. At high concentrations (that vary from one chemical species to another) there is deviation from the Beer–Lambert Law. Use your plotted data as a guide for identifying the point at which deviation occurs. If you do not observe deviation from linearity, your concentrations could be in the correct range. Once you know the amount of copper in the coin, consider how you might determine the amount of zinc.

When you consider waste disposal issues, recall that you added excess nitric acid to dissolve the coins. Consider what you might add to neutralize the acid in all your solutions. (Don't forget the solution in your cuvette.) In addition, think about how you might separate the metal ions from the water and make your recommendations in your report.

Report

The report must be typed and grammatically correct. Include a cost analysis for preparation of pennies based on raw material prices. Price information about copper and zinc can be found in many newspapers and on the Internet. Assume that manufacturing costs are the same regardless of penny composition.

> **Caution:**
> Wear your goggles at all times. Nitric acid can severely damage skin and eyes. NO_2 is a toxic gas — all reactions must be carried out in a fume hood.

Investigation 18

How Much Copper Is in the Coin? Team Contribution Form

Identify the percent contribution each team member made in this investigation, listing your name first.

Name: Percent Contribution:

_____   _____

_____   _____

_____   _____

_____   _____

Describe your contribution to the completion of the investigation and the writing of the report.

Return this sheet to your lab instructor.

Investigation 19
Which Iron Compound Is It?

Introduction

An old chemical stockroom belonging to a retired professor contains a shelf labeled "iron compounds." On this shelf are four bottles and an equal number of labels that fell off the bottles. A new faculty member needs the retired professor's lab space so old materials must be discarded. To dispose of the chemicals properly, they must be identified and labeled. The college has asked your group to develop a method to determine which label belongs to which bottle. The set of labels contained the following formulas of iron(III) compounds: $Fe_2(SO_4)_3 \cdot 5H_2O$, $FeCl_3 \cdot 6H_2O$, $Fe(NO_3)_3 \cdot 9H_2O$, $Fe(NH_4)(SO_4)_2 \cdot 12H_2O$.

Goals

As you complete this investigation you will:
1. Develop a method for determining the amount of iron(III) in an iron compound.
2. Use this method to determine the amount of iron(III) in the unknown compounds.
3. Use the results to identify the unknown iron compounds.

Materials

$Fe(NH_4)(SO_4)_2 \cdot 12H_2O$ (solid)
Solutions of the unknown iron compounds*
6 M HNO_3(aq)
1.0 M KNCS(aq)
Device for measuring light absorbance

Cuvettes
Balances
Tissues (preferably lint free)
Volume measuring equipment
Other supplies by request

*Obtain exact concentration, in mass of solid per liter of solution, from label.

Getting Started

A common color test for the presence of iron(III) in a solution involves the formation of the blood-red isothiocyanatoiron(III) and other similarly colored complex ions:

$$Fe^{3+}(aq) + NCS^-(aq) \rightarrow FeNCS^{2+}(aq)$$

This complex ion can be used as a light-absorbing species that relates to the amount of iron(III) present in a solution. As you design your experiment there are several factors you need to consider. Determine how much thiocyanate must be added to get a red color that is proportional to the iron(III) concentration. Can the thiocyanate concentration be varied or must it be held constant in different solutions? Does the iron have a color in the absence of the thiocyanate ion? Note that iron(III) tends to hydrolyze in neutral solution, forming a hydroxoiron(III) ion:

$$Fe^{3+}(aq) + H_2O(l) \rightarrow FeOH^{2+}(aq) + H^+(aq)$$

This reaction can be suppressed by the addition of some nitric acid. How much nitric acid is needed? Must the amount be kept constant in different solutions?

For information about the basics of spectroscopy, see Appendix B of this manual. In Appendix B the quantity, percent transmittance (%T), is described as a relationship between the intensity of light before and after passing through a solution. Another quantity known as absorbance (A) is related to %T by the equation:

A = log(100/%T)

Absorbance is related to the concentration of a light-absorbing species by the Beer-Lambert Law, **A = εbC,** where ε is a proportionality constant called the molar absorptivity that is specific for the absorbing species, **b** is the thickness of the solution in the light path (usually constant for an experiment), and **C** is the molarity of the absorbing species in the solution. For a given light-absorbing species, a plot of absorbance versus concentration should give a straight line. This straight-line relationship can be used to determine the concentration in the unknown solution.

Use the instructions in Appendix B for calibrating the spectrometer or colorimeter. (If these are unavailable you can use the non-instrumental technique described in Appendix B.) You will have to determine the wavelength at which to run your experiments. Solid iron(III) ammonium sulfate is available for preparation of your known solutions. You will have to determine the range of concentrations and develop procedures for using the iron(III) ammonium sulfate to make those solutions (at least seven). Collect absorbance data for each concentration you prepare. Construct a plot of your data that you can use to determine the concentration in an unknown solution. At high concentrations (that vary from one chemical species to another) there is deviation from the Beer-Lambert Law. Use your plotted data as a guide for identifying the point at which deviation occurs. If you do not observe deviation from linearity, your concentrations could be in the correct range.

Report

Submit a report in the required scientific fashion to your instructor. In addition to the usual items, your discussion should address relationships you identified between the concentration of the iron(III) in a solution and the color intensity. In addition, you should compare the measured percent iron in the unknown compounds with the expected percent iron for the compounds on the labels. This will allow you to determine the identity of the compounds in the bottles.

Caution:
Wear your goggles at all times. Nitric acid can severely damage skin and eyes.

Which Iron Compound Is It? Team Contribution Report Form

Identify the percent contribution each team member made in this investigation, listing your name first.

Name: Percent Contribution:

_____   _____

_____   _____

_____   _____

Describe your contribution to the completion of this investigation and the writing of the report.

Return this sheet to your lab instructor.

Investigation 20
Should We Mine This Ore?

Introduction

Copper and other ores arise from geothermal activity underground. Metal sulfides undergo oxidation reactions in the presence of water to form underground sulfuric acid. As a result, aqueous metal sulfates can form in the presence of sulfuric acid. The dissolved copper sulfate (and other metal sulfates) then seep downward through cracks in the surrounding rock. When the dissolved minerals reach the water table, the sulfuric acid is diluted, which causes the metal ions to precipitate out of solution. Copper is deposited as a mineral such as chrysocolla, $CuSiO_3 \cdot 2H_2O$. Geologists locate ore deposits by assessing characteristic geological features, then they sink drill holes to bring up underground rock samples. Chemists analyze the rock samples to see if the underground deposits are economically feasible sources of the desired metal. Once a good ore deposit is located, the mining companies remove the ore from deep underground. The ore is crushed into small pieces and acid is poured onto piles of the crushed ore. Dissolved copper then collects in pools at the bottom of the piles of ore as the acid soaks through the ore.

Geologists have identified several drill locations at sites that might contain copper. They have sent some crushed rocks on which you are to conduct copper analyses.

Goals

As you complete this investigation you will:
1. Develop a method for determining the concentration of copper(II) ion solution.
2. Determine how much copper(II) is present in an ore sample.
3. Report your findings in the usual scientific manner.

Materials

Processed copper ore
$Cu(NO_3)_2 \cdot 3H_2O(s)$
$6\ M\ HNO_3(aq)$
$6\ M\ HCl(aq)$
$6\ M\ CH_3CO_2H(aq)$
$6\ M\ NH_3(aq)$
Filter paper

Device for measuring light absorbance
Cuvettes
Tissues (preferably lint free)
Balances
Volume measuring equipment
Other supplies by request

Getting Started

The copper-containing compound in the rock can dissolve in acid to produce a blue solution. You will have to determine which acid to use. The reaction will become very slow if there is not enough acid present. Consider what might be a reasonable excess of acid to make the reaction proceed faster. Copper ores tend to have only a small amount of copper. Since the solution is likely to be only slightly colored, it may be necessary to enhance the color intensity by adding excess aqueous ammonia. How will you know that you have excess ammonia? A deep blue complex ion, $Cu(NH_3)_4^{2+}$, is formed in an amount equal to the amount of Cu^{2+} present initially:

$$Cu^{2+}(aq) + 4NH_3(aq) \rightarrow Cu(NH_3)_4^{2+}(aq)$$

For information about the basics of spectroscopy, see Appendix B of this manual. In Appendix B the quantity, percent transmittance (%T), is described. Another quantity known as absorbance (A) is related to %T by the equation:

$$A = \log(100/\%T)$$

Absorbance is related to the concentration of a light-absorbing species by the Beer-Lambert Law, $A = \varepsilon bC$, where ε is a proportionality constant called the molar absorptivity that is specific for the absorbing species, **b** is the thickness of the solution in the light path (usually constant for an experiment), and **C** is the molarity of the absorbing species. For a given light–absorbing species, a plot of absorbance versus concentration should give a straight line. This straight–line relationship can be used to determine the concentration of an unknown solution.

Use the instructions in Appendix B for calibrating the spectrometer or colorimeter. (If these are unavailable you can use the non-instrumental technique described in Appendix B.) You will have to determine the wavelength at which to run your experiments. Solid copper(II) nitrate and ammonia are available for preparation of your known solutions. You will have to determine the range of concentrations and develop procedures for using the copper(II) nitrate and ammonia to make those solutions (at least seven). Collect absorbance data for each concentration you prepare. Construct a plot of your data that you can use to determine the concentration in an unknown solution. At high concentrations (that vary from one chemical species to another) there is deviation from the Beer-Lambert Law. Use your plotted data as a guide for identifying the point at which deviation occurs. Once you have a standard plot you will use this information to determine the amount of copper in the ore sample.

Report

Your report should be written in the usual scientific fashion. Your discussion should include your recommendations to geologists regarding the drill sites.

Caution:
Wear your goggles at all times. You are working with strong acids that can severely damage skin and eyes. You might generate fumes that are toxic. All reactions must be carried out in a fume hood.

Should We Mine This Ore? **Team Contribution Form**

Identify the percent contribution each team member made in this investigation, listing your name first.

Name: Percent Contribution:

_____   _____

_____   _____

_____   _____

_____   _____

Describe your contribution to the completion of the investigation and the writing of the report.

Return this sheet to your lab instructor.

Investigation 21
What Causes Intermolecular Attractions?

Introduction

A company has produced a toy that contains layers of liquids that can be shaken to give interesting patterns, but then separate back into the original layers. It is trying to obtain a patent for its toy, but has been challenged by another toy company which makes a similar product. The patent lawyers want your group members to serve as expert witnesses to explain the nature of the intermolecular attractions between the different substances in the toy. They want you to use molecular models to explain why molecules might attract or repel one another to form layers. The company also wants you to provide explanations that can be understood by a judge and jury who might not have studied chemistry.

Goals

As you complete this investigation you will:
1. Use a molecular model kit to build models that illustrate the structures of various molecules.
2. Use the models to decide the nature of the attractive forces between these molecules.
3. Decide which substances could exist together in a layer.
4. Explain the behavior of the layers by using the molecular models.

Materials

Molecular model building kit or materials that can be used to construct molecules

Getting Started

Start by building models of C_2H_6, C_2H_4, and C_2H_2. Atoms should be arranged so they are as far away from each other as possible. (See your VSEPR in your textbook for additional information about this topic.) Use these models to explain differences in atom arrangement and bond angles. Describe the bond angles in these molecules. How do the shapes of these molecules differ?

Using models of CH_4, NH_3, and H_2O develop an explanation for describing shapes of molecules, polarity, and intermolecular forces. Compare the arrangement of the hydrogens around the central atom for each of the structures. What is the arrangement of atoms and valence electron pairs around the central atom? What are the approximate bond angles? What are the shapes of the molecules? Which molecules are polar and which are nonpolar? Which of the compounds has the highest boiling point? Which molecule has the strongest intermolecular forces? You might consider building multiple structures of the same molecule to consider how multiple molecules interact.

Investigation 21

To develop your explanations about intermolecular forces further, use models to explain the differences in boiling points for the compounds listed in Table 1. The chemical formulas listed in the table are shorthand notations for the arrangement of atoms along the carbon chain. The figure below shows the attachment of hydrogens around the backbone structure for the first molecule listed in Table 1. Note that the structural formula as shown is written in two dimensions and does not represent the three-dimensional arrangement of atoms needed to explain the properties of the molecule.

$$CH_3CH_2OH \quad = \quad \begin{array}{c} H \ \ H \\ | \ \ | \\ H-C-C-O-H \\ | \ \ | \\ H \ \ H \end{array}$$

Consider each of the compounds in the liquid state. How would you describe the forces of attraction between the molecules?

Table 1

Name	Condensed Formula	Boiling Point
Ethanol	CH_3CH_2OH	78.5 °C
Propanol	$CH_3CH_2CH_2OH$	97.4 °C
Butanol	$CH_3CH_2CH_2CH_2OH$	117.2 °C

The toy manufacturer mixes carbon tetrachloride (CCl_4), water, and the chemicals shown in Table 2 to make the liquid that is in two layers. Based on your models, decide what types of intermolecular forces could be experienced by each of these substances. Use the models to determine which chemicals compose each layer that makes up the liquid within the toy. The toy manufacturer wants one of the layers to have color so it is adding iodine to make a layer that is purple. In which layer can the iodine be found?

Table 2

Cyclohexane C_6H_{12}	Diethyl ether $CH_3CH_2OCH_2CH_3$	Ethylene glycol $HOCH_2CH_2OH$
$\begin{array}{c} H_2C-CH_2 \\ / \ \ \ \ \ \ \ \ \ \backslash \\ H_2C \ \ \ \ \ \ \ CH_2 \\ \backslash \ \ \ \ \ \ \ \ \ / \\ H_2C-CH_2 \end{array}$	$\begin{array}{c} H \ \ H \ \ \ \ H \ \ H \\ | \ \ | \ \ \ \ | \ \ | \\ H-C-C-O-C-C-H \\ | \ \ | \ \ \ \ | \ \ | \\ H \ \ H \ \ \ \ H \ \ H \end{array}$	$\begin{array}{c} H \ \ H \\ | \ \ | \\ HO-C-C-OH \\ | \ \ | \\ H \ \ H \end{array}$

Report

Submit a report that adequately describes the structures you were asked to build in this investigation. Discuss the nature of the intermolecular forces and boiling points for the pure substances. Also, discuss the composition of the layers in the toy and the intermolecular forces between the compounds that make up the layers.

What Causes Intermolecular Attractions? **Team Contribution Form**

Identify the percent contribution each team member made in this investigation, listing your name first.

Name: Percent Contribution:

_____   _____

_____   _____

_____   _____

_____   _____

Describe your contribution to the completion of the experiment and the writing of the report.

Return this sheet to your lab instructor.

Investigation 22
What are the Structures of Some Alloys?

Introduction

A company that manufactures surgical devices for scoliosis, a disease that causes three dimensional deformation of the spine, is in search of new metal alloys to stabilize the spine after corrective surgery. The company currently produces a NiTi alloy (1:1 atom ratio), called memory metal because the metal can change shape when a stress is placed on it and then regain its former shape. Memory metals are ideal for this type of surgery because they allow the movement of the spine while providing continued support. The problem is that Ni can also be toxic or cause allergic responses in humans. The company would like your team to investigate the structures of beryllium and gold alloys with a 1:1 atom ratio to see if they could be used to make devices that will not cause potentially harmful side-effects. The company is willing to financially support the research, but requires a funding proposal to be submitted based on information that you will gather during this investigation. After consideration of the submitted funding proposals, one project will be selected and manufacturing can begin.

Goals

As you complete this investigation you will:
1. Compare the structures of simple (primitive) cubic, body-centered cubic, and face-centered cubic unit cells, identifying atom arrangement and packing efficiency of each type.
2. Calculate the atomic radius of Au given the edge length and type of unit cell.
3. Develop two possible structures for beryllium-gold alloys.
4. Write a proposal to the company that shows justification for the BeAu alloy structures.

Materials

Crystal model building kit
Paper unit cell models (optional)

Getting Started

To start, you should build models of the three types of cubic structures (refer to model kit manual). You can use information gained from these models to compare the structures of simple cubic, body-centered cubic, and face-centered cubic unit cells. You might notice that some atoms are touching each other and some are not. The number of atoms that touches a central atom is called the *coordination number*. Also notice that there are holes created by the arrangement of atoms. You should build the extended structures to see some of the unit cell characteristics. From the extended structure, determine where one unit cell ends and the next begins. Also,

Investigation 22

determine the coordination number for each type of unit cell. Are the holes in each structure the same? If not, describe the arrangement of atoms around the holes.

The amount of open space in a unit cell is related to the packing efficiency of the atoms and, thus, the density of species that assume those structures in the solid form. For each structure determine the percent packing efficiency, which is a function of the amount space occupied be atoms within the unit cell and the total volume of the cell. (If you define the dimensions of the cell in terms of the radius of the atoms, you do not need the dimensions to calculate packing efficiency.) Compare the packing efficiency for each structure. Assuming a spherical atom shape, the volume of a sphere is given by the equation:

$$V_{Sphere} = \frac{4}{3}\pi r^3$$

Now that you have seen basic unit cell structures, you can consider structures that involve different elements. An alloy is a homogenous mixture of two or more metals. The metals combine at the atomic level in one of two ways. One way is replacing one atom with another, such as substituting a Ni atom for a Ti atom. The alternative is for an atom to occupy some or all of the holes in the structure. For instance a Ni atom could occupy the cubic hole in a simple cubic arrangement of Ti atoms.

The atomic radius of beryllium is 90 picometers. Gold is known to exist in a face-centered cubic arrangement. Crystallographic studies indicate that the edge length of a gold unit cell is 506.288 picometers. You will need to calculate the atomic radius of gold. Using the ratios of the sizes of these atoms, you should choose the appropriately sized spheres from the kit to build your team's proposed 1:1 BeAu alloy structures.

Report

Your group will write this report as a funding proposal rather than your normal lab report format. It needs to be typed and grammatically correct. The goal of this proposal is to convince the alloy manufacturer that your team's BeAu structures are worth further investigation and investment. You must provide adequate evidence that your team not only understands the basics of crystal structures, but also that the BeAu structures are feasible. A suitable proposal should include the following:

- a description of the types of cubic unit cells (number of the atoms per unit cell, the number and types of holes per unit cell, the coordination number, and packing efficiency)
- a description of possible structures for the BeAu alloys using appropriate terminology and sketches
- reasons why one proposed alloy may be better than the other
- the potential benefits/risks of using beryllium and gold alloys in the human body

Investigation 22

What Are the Structures of Some Alloys? Team Contribution Form

Identify the percent contribution each team member made in this investigation, listing your name first.

Name: Percent Contribution:

_____   _____

_____   _____

_____   _____

_____   _____

Describe your contribution to the completion of the investigation and the writing of the report.

Return this sheet to your lab instructor.

Investigation 23
How Is LED Light Color Related to Composition?

Introduction

An electronics firm has sent light emitting diodes of various composition for your team to test. Instead of the usual red LED light that appears on many indicator electronic devices, the firm is considering other colors. The company wants your group to identify the colors of the light-emitting diodes and relate the observed color to chemical composition. The company also wants you to determine relationships among composition, observed diode color, measured voltage, and energy of the light given off by the diode.

Goals

As you complete this investigation you will:
1. Identify the color of various LEDs.
2. Estimate energy of the light emitted from the LEDs.
3. Relate the observed differences in LED colors to chemical composition.
4. Relate the observed differences in LED voltages to chemical composition.

Materials

Light-emitting diodes of various composition*
Circuit board with 1 kΩ, 10 kΩ, 50 kΩ, and 1 MΩ resistors
9 V batteries
Device to measure voltage
  Note the chemical composition of each LED.

Getting Started

Identify the color of each diode by inserting it into an appropriate socket on the circuit board. How does the light intensity vary with resistance? Based on your data you should be able to estimate the energy associated with the light emitted from the LEDs. You may have to consult your textbook to find information about energy and light. You should also determine the relationship between composition and the observed color.

Determine the voltage across each LED. (See Appendix F of your lab manual for information about collecting voltage data.) Explain the differences among voltages observed for the various diodes. What is the relationship between the observed voltage and the energy of the light emitted from each LED?

Investigation 23

Report

Your report for this investigation should be in the form of a letter to the electronics firm. It should include a discussion of relationships between observed light emitted and chemical composition. You should also discuss your observations about the voltages measured across the LEDs. Include your recommendations for selection of LED colors in electronic devices.

Investigation 23

How Is LED Light Color Related to Composition? Team Contribution Form

Identify the percent contribution each team member made in this investigation, listing your name first.

Name: Percent Contribution:

_____   _____

_____   _____

_____   _____

_____   _____

Describe your contribution to the completion of the investigation and the writing of the report.

Return this sheet to your lab instructor.

Investigation 24
Are Pollutant Gases Harmful to Plant Life?

Introduction

Government regulation of industrial activities is an emotional topic among many segments of our society. For example, politicians and environmentalists are sometimes at odds about scientific evidence claiming the detrimental effects of low pollutant-gas concentration. Indeed, the contribution of manufacturing activities to pollution is well documented. Sulfur dioxide (SO_2) and nitrogen dioxide (NO_2) are common pollutant gases and contribute to acid rain. However, certain segments of society are skeptical about the existing evidence that shows that these gases are dangerous to human and plant life. In this investigation your team of researchers has been asked to provide evidence to support or refute the claim that sulfur dioxide and nitrogen dioxide are harmful to plant life.

Goals

As you complete this investigation you will:
1. Synthesize three gases, CO_2, SO_2 and NO_2, for testing.
2. Test the gases' pH and effect on various organisms.
3. Develop a technique for disposal of the waste gases.
4. Report your results and their implications for environmental regulation.

Materials

Well plates	Sodium bicarbonate
60 mL LuerLOK syringes	Iron(III) chloride
Plastic caps	3 M sulfuric acid
Resealable plastic bags	6 M hydrochloric acid
Disposable pipets	1 M sodium hydroxide
Microscopes and slides	6% hydrogen peroxide
Universal indicator	Blepharisma colony
pH reference solutions (pH 3 to 8)	Chrysanthemum flowers
Sodium bisulfite	Moss sample
Potassium nitrite	

Getting Started

Your task in this investigation is to design an experiment for producing the gases, testing their pH, and observing their behavior on Blepharisma, a pond-dwelling, one-celled organism. You will test the effect of the gases on a chrysanthemum (or other flower) and a moss. Perfect your synthetic technique by preparing carbon dioxide first.

Investigation 24

The gases you will study are not readily available, so your team will have to produce them from reagents provided. You should begin by developing a technique for synthesizing CO_2. You can produce this gas from sodium bicarbonate and hydrochloric acid. Sodium chloride and water are by-products of the reaction. How might you generate CO_2 gas without performing a chemical synthesis in the laboratory? Why is laboratory synthesis more appropriate for scientific testing?

After perfecting your CO_2 synthesis technique, you should determine methods for producing SO_2 and NO_2. **Be careful: SO_2, NO, and NO_2 are toxic gases. Run all experiments in a sealed environment.** For the synthesis of SO_2, sodium bisulfite can react with hydrochloric acid to produce the gas. (*If you have a sensitivity to bisulfite, inform your instructor.*) Sodium chloride and water are by-products of the reaction. In designing your synthesis, keep in mind the danger of working with strong acids and the generation of toxic gases.

Preparation of NO_2 will be a little more complex because you will synthesize it from two other gases, NO and O_2. You can produce NO_2 in a manner similar to your synthesis of SO_2. Potassium nitrite can react with sulfuric acid to generate NO_2. Potassium sulfate, potassium nitrate, and water are by-products of this reaction. Oxygen gas can be prepared from the decomposition of H_2O_2 catalyzed by $FeCl_3$. Develop a technique to mix the NO and O_2 gases without exposing anyone to the toxic gases. Alternatively, you can draw air into the reaction chamber. When combined, the NO gas and O_2 from the air will produce NO_2. Again, you should perfect ways of capturing the gas with no loss to the environment.

Once you have designed your method of gas production, consider how you will test each gases' pH and effect on the organisms. Blepharisma is a pink, oval, single-celled organism found in pond water. The organism is observable with a 100X microscope. Consider how you might test the effects of the gases under environmental conditions (lower concentrations of gas).

In studying pH, you can set up reference solutions of known pH with universal indicator for comparison. (Refer to Appendix C for more information about measuring pH.) Is there a pattern in the acidity or basicity of the gases? Are there environmental consequences for this pattern?

Report

Submit your report in the usual scientific manner. Include chemical equations for all reactions. Thoroughly describe your results and their environmental implications. What are common levels of the pollutant gases you tested? How would your findings differ if these levels were tested?

Caution:
Wear your goggles at all times. Sulfuric acid is very dangerous.
SO_2 and NO_2 are toxic gases. Run all reactions in a fume hood.

Are Pollutant Gases Harmful to Plant Life? Team Contribution Report Form

Identify the percent contribution each team member made in this investigation, listing your name first.

Name: Percent Contribution:

_____   _____

_____   _____

_____   _____

_____   _____

Describe your contribution to the completion of this investigation and the writing of the report.

Return this sheet to your lab instructor.

Investigation 25
What is the Molar Mass of Mars Ice Gas?

Introduction

Photographs taken by the Mars Global Surveyor indicate that the permanent ice on the Mars north pole and south pole are different. It has been determined from remote spectroscopic studies that the Mars atmosphere consists primarily of carbon dioxide (95.3%) and nitrogen (2.7%), along with small amounts of argon, oxygen, carbon monoxide, water, and nitrogen oxide. Various theories propose that the ice on Mars could consist of carbon dioxide, water, or methane hydrate (formed by volcanic activity). Since methane, water, and carbon dioxide could be used as basic elements of human habitation, it is important for future exploration and possible colonization of this planet that the identity of the ices on the poles be determined. The temperature of the north pole is sufficiently high that it is suspected that its ice must be water ice. However, at the south pole, temperatures are low enough that the ice could be carbon dioxide ice or methane ice. The National Space and Aeronautics Administration (NASA) sent a Mars Explorer to the south pole to collect a sample and return it to Earth in a refrigerated container. This ice does not melt to give a liquid, but vaporizes fairly readily, so it is probably either carbon dioxide or methane, but may be some other unsuspected compound. To determine the identity of this ice, NASA has asked you to develop a method of measuring its molar mass.

Goals

As you complete this investigation you will:
1. Develop a method for measuring the molar mass of a gas derived from vaporization of a solid.
2. Apply the gas laws to this problem.
3. Use the measured molar mass to predict the chemical identity of the solid.

Materials

Sample of Mars ice
Tongs
Erlenmeyer flask, 250 mL
Rubber stopper, one-hole, #6
Distilled water
Balances
Barometer or other pressure-measuring device
Other supplies by request

Investigation 25

Getting Started

The solid has been kept at quite cold temperatures, so you should not handle it directly with your hands. It is also sufficiently cold that it might condense water vapor from the air onto its surface. When you select a sample for your experiment, you should wipe it with a clean towel first. Because the vaporizing gas can generate pressures high enough to cause a glass flask to explode, do not stopper the flask tightly. As you plan your procedure, you should consider the following factors. If you are measuring the mass of a flask, do your fingerprints on the glass make a difference? How will you measure the volume of gas in the flask? How will you measure the pressure of the gas in the flask? Is an empty flask actually empty? What does it contain? You might find use for the following equation for the density of air in units of g/mL:

$$d_{air} = 0.3535 \frac{P}{T}$$

In this equation, *P* is pressure in atmospheres and *T* is temperature in Kelvins.

Report

Submit a report to NASA that describes the method you developed for measuring molar mass and the results obtained for Mars ice. Include a prediction of the identity of Mars ice based on the value of the molar mass. Suggest other experiments that might be done to identify this ice. Suggest some chemical processes that could be used to convert the ice compound into other useful compounds.

> **Caution:**
> **Wear your goggles at all times. You are working with materials of unknown composition so you must take precautions.**

What is the Molar Mass of Mars Ice Gas? Team Contribution Form

Identify the percent contribution each team member made in this investigation listing your name first.

Group: Percent Contribution:

_____   _____

_____   _____

_____   _____

_____   _____

Describe your contribution to the completion of the investigation and the writing of the report.

Return this sheet to your lab instructor.

Investigation 26
How Much Gas Is Produced?

Introduction

Chemical manufacturing facilities often have to consider the consequences of gas production when running chemical reactions in bulk. The amount of gas produced can be predicted for many types of chemical reactions from stoichiometry and gas law relationships. However, the amount of product often differs from the predicted quantity as determined from calculations that assume ideal behavior.

A chemical engineering firm has asked your team to investigate gas production from various chemical reactions. The firm wants your team to measure the amounts of gaseous product obtained from two or more reactions and compare them to the amount that is predicted on the basis of the law of conservation of mass and ideal gas behavior. The company wants to compare the yields obtained by your group in the lab with data it has collected from the manufacturing facility. Where there are deviations from expected amounts, the company needs explanation.

Goals

As you complete this investigation you will:
1. Develop a method for measuring the amount of gaseous product obtained from various reactions.
2. Apply the gas laws to this problem.
3. Compare predicted and measured yields for the gases you produce.
4. Consider the implications of gas production on the reaction of 1000 pounds of solid raw material from your reactions.

Materials

1 M HCl(aq)	Manometer, pressure sensor, or apparatus to collect gas by water displacement
1 cm strips of Mg ribbon	
$CaCO_3(s)$	25 mL syringe with connectors
$Na_2CO_3(s)$	Tygon tubing
Balances	50 mL flask with one-hole stopper fitted with dropper tube
Thread	125 mL flask with one-hole stopper fitted with dropper tube
Tissues	Volume measuring equipment
Thermometer	Other supplies by request

Getting Started

Investigation 26

Your instructor will assign two or more of the solids (Mg and $CaCO_3$ or Na_2CO_3). You will add hydrochloric acid to your assigned solids and collect the gases that the reactions produce. **Note that the reaction of magnesium with hydrochloric acid produces flammable hydrogen gas.** Before you begin collecting gas from your reactions you should become familiar with the apparatus you will use to measure the amount of gas produced. (If you use a pressure sensor or manometer, you may want to verify that the apparatus is working properly. What tests could you run to make this determination?)

Assemble an apparatus that will allow you to mix the solid with **excess** hydrochloric acid in a closed container that is attached to a pressure sensor or manometer. Alternatively, you could collect the gas produced by displacing water held in a tube or bottle inverted in a trough of water. (If you collect your gases by water displacement you must account for the vapor pressure of water.) As you design your experiment consider the following questions. Why should you use excess acid? How much excess? Why should you wait to mix the reactants until after the container is sealed?

If you use a pressure sensor to collect pressure data, note that the seal between the flask and the stopper may break if the pressure increase becomes as great as 0.5 atm. Your group should do some preliminary calculations to show that your system will not exceed a total pressure of 1.5 atm. Make sure that your procedure does not allow any liquids to enter the opening of the pressure sensor. Consult the instructions in Appendix E for guidance in using your equipment to measure pressure. Consider other variables you will need to measure to determine the amount of gas produced.

Compare the amount of gas produced by the reaction to the amount that is predicted from stoichiometric and ideal gas calculations. Is there good agreement? If not, and you suspect some aspect of your method is at fault, revise your method and try again. If errors remain after making any reasonable modifications, propose explanations for the discrepancies. You might find that one of the reactions produces results that are close to the theoretical quantities expected while other reactions do not. Is there an explanation for these differences?

Report

The report must be typed and grammatically correct. It may be returned to you for corrections if it is not acceptable. Thoroughly discuss your results and include chemical equations for the reactions. Your discussion should include explanations for deviations from expected results.

> **Caution:**
> **Wear your goggles at all times. HCl is a strong acid. Hydrogen gas is flammable. You are generating pressure in a sealed flask. Inspect all equipment for cracks.**

How Much Gas Is Produced? **Team Contribution Form**

Identify the percent contribution each team member made in this investigation, listing your name first.

Name: Percent Contribution:

_____   _____

_____   _____

_____   _____

_____   _____

Describe your contribution to the completion of the investigation and the writing of the report.

Return this sheet to your lab instructor.

Investigation 27
Which Alcohols Are in the Barrels?

Introduction

Suppose you work for a company that produces methyl alcohol, ethyl alcohol, and isopropyl alcohol. Last night the production people forgot to label one of the barrels. The shipping department needs to know the identity of the alcohol so it can go on the correct truck. Your team has been called to identify the alcohol in the barrel. The only chemical available to you besides your unknown alcohol is tertiary butyl alcohol, which freezes at close to room temperature. All of the alcohols produced by your company dissolve in tertiary butyl alcohol.

Goals

As you complete this investigation, you will:

1. Design an experiment based on colligative properties that will determine the molar mass of the unknown alcohol.
2. Use the experimental molar mass to determine which alcohol is in the barrel.
3. Report your findings in the usual scientific manner.

Materials

2 or 3 unknown alcohols
Tertiary butyl alcohol
Thermometer
Large test tubes
One–hole stopper for test tube

Ice
Warm water
Large beakers
Balances
Other supplies by request

Getting Started

Since tertiary butyl alcohol (*t*-butyl alcohol) freezes at close to room temperature, you should be able to measure its freezing point. Be aware that a liquid can temporarily cool below its freezing point while the particles are getting organized. This phenomenon is known as *supercooling*. Once the freezing begins, the temperature will settle at the freezing point and freezing will be observed. Also note that an unmixed liquid can temporarily have regions of different temperatures, so continuous light stirring is recommended. Compare your experimental freezing point of *t*-butyl alcohol to that found in the literature.

To begin your experiments to determine the identity of the unknown alcohols, you might observe what happens to the freezing point upon addition of one of the unknowns to some *t*-butyl alcohol. If you add an equal mass of your unknown alcohol to a sample of the *t*-butyl alcohol, does the freezing point stay the same? Can the changes in the freezing point be used to determine the molar mass of the unknown alcohols? If you found that the freezing point change when the unknown was added to *t*-butyl alcohol, the section on colligative properties in your textbook might be of use. The change in freezing point of a solution is given by the equation:

$$\Delta T_f = K_f m$$

In this equation ΔT_f is the change in freezing point, m is the molality of the solution, K_f and is a constant known as the freezing point depression constant. This constant is specific to a given solvent. For *t*-butyl alcohol the freezing point depression constant is $8.3\,^\circ C/m$. How can the molar mass of a solute be determined from the molality of a solution containing this solute?

When identifying your alcohols, note that methyl alcohol is also known as methanol or wood alcohol, ethyl alcohol is also known as ethanol, and isopropyl alcohol is also known as 2-propanol.

Report

The report must be typed and grammatically correct. It may be returned to you for corrections if it is not acceptable. Thoroughly discuss your results and include the identity of the unknown alcohols. If directed by your instructor, compare you results to those obtained by other groups who had the same unknowns.

Caution:
Wear your goggles at all times. The alcohols in this experiment are not for human consumption. Methanol is toxic. Because the alcohols are flammable, no open flames are allowed in the lab.

Investigation 27

Which Alcohols Are in the Barrels? Team Contribution Form

Identify the percent contribution each team member made in this investigation, listing your name first.

Name: Percent Contribution:

_____   _____

_____   _____

_____   _____

Describe your contribution to the completion of the investigation and the writing of the report.

Return this sheet to your lab instructor.

Investigation 28
How Is Heat of Combustion Measured Indirectly?

Introduction

A chemical reaction is usually represented by a chemical equation, which summarizes the relative amounts of the various substances that appear as reactants and products. A complete summary of the reaction would also include the enthalpy or heat change accompanying the reaction. Values of such enthalpy changes have been determined for many reactions. For example,

$$H_2(g) + \tfrac{1}{2} O_2(g) \rightarrow H_2O(l) \quad \Delta H = -285.8 \text{ kJ}$$

For the above reaction, 285.8 kJ of energy is released upon reaction of one mole of H_2 gas with ½ mole of O_2 gas to produce one mole of H_2O liquid, as described by the coefficients in the equation.

Suppose a ceramics company that uses magnesium oxide as a thermal insulator needs to know the heat of combustion of Mg as described by the reaction:

$$Mg(s) + \tfrac{1}{2} O_2(g) \rightarrow MgO(s) \quad \Delta H = ?$$

Unfortunately, the heat change associated with this reaction cannot be measured directly with the equipment available in the lab because it is dangerous to burn magnesium. An alternative approach involving other chemical reactions will have to be used. The ceramics company has contracted for your group to design and carry out a procedure that will indirectly determine the heat of combustion of magnesium metal and report your results.

Goals

As you complete this investigation you will:
1. Design a procedure to measure heat changes associated with a chemical reaction.
2. Apply Hess's law to determine heat changes that cannot be measured directly.
3. Summarize your findings in a written report.

Materials

1.0 M HCl(aq)
Mg ribbon
MgO(s)
Styrofoam cups
Thermometer or other temperature probe
Other supplies by request

Investigation 28

Getting Started

A common method for determining the heat associated with a reaction occuring in aqueous solution involves the use of a Styrofoam cup calorimeter. For example, if NaOH(*aq*) and HCl(*aq*) were mixed in the cup, and a temperature change was measured, then the heat associated with the reaction could be calculated by knowing the mass of the solution involved and the specific heat of the aqueous solution, which can be assumed to be 4.184 J/g·°C. This assumption is not strictly accurate, but it will work adequately for the solution concentrations used in this investigation. Note that for a chemical reaction run in solution, the heat evolved (or absorbed) by the reaction will be absorbed (or released) by the solution, as indicated by the first law of thermodynamics. With the reagents provided, could you determine various ΔH values for different reactions, that when combined in some way allow you to calculate ΔH for the combustion of magnesium to give magnesium oxide? (See Hess's law in your for information about how this might be done.)

You should observe a reaction before you measure the value of ΔH to make sure that you have selected appropriate conditions to make a heat measurement. For example, how will you handle your calculations if the solid for one of your reactions is not the limiting reactant? Note that ΔH for a given reaction must correspond to the mole quantities described by the coefficients in the chemical equation. You will not be running your reactions using these quantities so you will have to make an adjustment to the heat change determined in your experiments. How will you make this adjustment?

The following thermochemical information may be of some use:

$$H_2(g) + \tfrac{1}{2}O_2(g) \rightarrow H_2O(l) \qquad \Delta H = -285.8 \text{ kJ}$$

Report

The report must be typed and grammatically correct. It may be returned to you for corrections if it is not acceptable. Your report should include all pertinent data collected to determine the heat of combustion of magnesium metal. The ΔH value should be reported in kJ/mol. The theoretical value of ΔH for this reaction can be found in your textook. How does your experimental value compare to the theoretical value?

Caution:
Wear your goggles at all times. HCl is a strong acid. Hydrogen gas is flammable. Do not burn the magnesium.

How Is Heat of Combustion Measured Indirectly? **Team Contribution Form**

Identify the percent contribution each team member made in this investigation, listing your name first.

Name: Percent Contribution:

_____   _____

_____   _____

_____   _____

_____   _____

Describe your contribution to the completion of the investigation and the writing of the report.

Return this sheet to your lab instructor.

Investigation 29
What Is the Rate Law?

Introduction

Factors that affect rates of chemical reactions include concentration, surface area, temperature, pressure, catalysis, solvent, and ionic medium. A kinetic study probes these factors and is carried out to determine the optimum conditions for a chemical reaction. Kinetic studies are also used to obtain information about the detailed manner in which a reaction occurs — that is, to determine the reaction mechanism.

Rate measurements are usually made under conditions such that all of the factors which might affect the rate are held constant, except one, so that the dependence of the rate on each variable is determined separately. Since molecules must come together to react, the rate of a reaction is expected to depend on the concentrations of reactants and possibly other substances. The dependence of the rate on concentrations is represented by a rate law, which is usually of the form:

$$\text{Rate} = k\,[A]^a[B]^b[C]^c$$

where k is the rate constant and the exponents are described as the order with respect to that particular substance.

Suppose your team is working in a chemical research facility to understand the kinetic nature of certain reactions. Specifically, your group will study the reaction of magnesium with HCl:

$$\text{Mg}(s) + 2\text{HCl}(aq) \rightarrow \text{H}_2(g) + \text{MgCl}_2(aq)$$

The research facility wants your group to monitor the rate of production of H_2 gas and determine the rate law for the reaction.

Goals

As you complete this investigation you will:
1. Design experiments that can measure the rate of production of H_2 gas as a function of time during the reaction of magnesium metal with an acid.
2. Determine the form of the rate law for this reaction.
3. Summarize your findings in a formal written report.

Investigation 29

Materials

1.0 M HCl(aq)	Pressure sensor, manometer, or apparatus to collect gas by water displacement
1 cm strips of Mg ribbon	Automated data collection device (CBL or MBL)
Balances	50 mL flask with one-hole stopper fitted with dropper tube
Small test tubes (3 mL)	125 mL flask with one-hole stopper fitted with dropper tube
Thread	Volume measuring equipment
Tissues	Other supplies by request
Thermometer	

Getting Started

Consider the reaction of interest in this investigation:

$$Mg(s) + 2HCl(aq) \rightarrow H_2(g) + MgCl_2(aq)$$

To begin your kinetic study, place a 1.0 cm strip of Mg metal into a flask containing 1.0 M HCl(*aq*), and observe the reaction. Based on your observations, design and carry out a series of experiments to determine the rate law and *k* value for the reaction. How will you describe the rate of the reaction in quantitative terms? You could assume initially that the rate depends on the HCl concentration and on the Mg surface area:

$$\text{Rate} = k \, (\text{Mg surface area})^x \, [\text{HCl}]^y$$

Your experiments should allow you to determine *k*, *x*, and *y* in the above expression. If a reaction rate is fairly constant initially, then the rate of that reaction can be determined by monitoring the molarity (or pressure or volume) of one of the reactants or products. Then you could plot the reactant or product molarity (or pressure) versus time, defining the slope of the line as the rate. If the reaction rate is not constant, then consult your textbook to see how to treat the data graphically to convert molarity (or pressure or volume) versus time data to rate data.

Report

The report must be typed and grammatically correct. It may be returned to you for corrections if it is not acceptable. Your report should include all pertinent data gathered and plotted to determine the rate law and rate constant value.

> **Caution:**
> Wear your goggles at all times. HCl is a strong acid. Hydrogen gas is flammable. Do not burn the magnesium.

What Is the Rate Law? **Team Contribution Form**

Identify the percent contribution each team member made in this investigation, listing your name first.

Name: Percent Contribution:

_____   _____

_____   _____

_____   _____

_____   _____

Describe your contribution to the completion of the investigation and the writing of the report.

Return this sheet to your lab instructor.

Investigation 30
How Fast Does the Crystal Violet Decolorize?

Introduction

A dye company recently had an entire production batch of crystal violet returned because it decolorized. Its customer uses a procedure in which a dilute crystal violet solution must retain its color in basic solution. The dye company wants your input on the feasibility of maintaining crystal violet color under basic conditions. If this is not feasible, the company wants a description of how fast the color is lost and a mathematical explanation for this behavior. Specifically, the company wants your interpretation of the kinetics of crystal violet decolorization.

Rate measurements are usually made under conditions such that all of the factors which might affect the rate are held constant, except one, so that the dependence of the rate on each variable is determined separately. Since molecules must come together to react, the rate of a reaction is expected to depend on the concentrations of reactants and possibly other substances. The dependence of the rate on concentrations is represented by a rate law, which is usually of the form:

$$\text{Rate} = k\,[A]^a[B]^b[C]^c$$

where k is the rate constant and the exponents are described as the order with respect to that particular substance.

Goals

As you complete this investigation you will:
1. Design an experiment that can measure the rate of decolorization of crystal violet as a function of time in basic solutions.
2. Determine the rate law for this reaction.
3. Report your findings in the usual scientific manner.

Materials

1.0×10^{-4} M crystal violet solution
0.10 M NaOH(aq)
1.0 M HCl(aq) (for cleaning glassware)
Spotplate

A device for measuring absorbance
Volume measuring equipment
Tissues (preferably lint free)
Other supplies by request

Investigation 30

Getting Started

Crystal violet will decolorize over a period of time when placed in basic solutions according to the equation:

[Structural equation showing crystal violet cation with three N(CH₃)₂-substituted phenyl groups attached to central C⁺ reacting with OH⁻ to form the corresponding carbinol with OH attached to the central carbon.]

The above equation can be abbreviated:

$$CV^+ + OH^- \rightarrow CVOH$$

Crystal violet decolorizes in the presence of hydroxide ion. You will determine the order of the crystal violet decolorization reaction with respect to the crystal violet concentration, $[CV^+]$. You may want to start by just observing the reaction. Place about 8 drops of water in a spot plate and add one drop of 1.0×10^{-4} M crystal violet solution. Now add one drop of 0.10 M NaOH(aq) (this represents a large excess of hydroxide ion). Did you notice the color fading? How long before there was a noticeable change? What are the initial concentrations of CV^+ and OH^-? Remember that each reactant is diluted when the two solutions are mixed, so the initial concentrations must be calculated.

Design and carry out an experiment to determine the order of this decolorization reaction with respect to $[CV^+]$ and with respect to $[OH^-]$, and the value of the rate constant, k. Is the value of either the order or the rate constant dependent on the concentration of hydroxide ion? Because crystal violet is a colored species a device that measures light absorbance may be of some use. The amount of light absorbed by a particular substance dissolved in water depends on the length of the light path through the solution, and on the concentration of the absorbing solute. The absorbance is related to concentration according to the Beer-Lambert law:

$$A = \varepsilon b C$$

where **A** is the absorbance, ε is a proportionality constant, called the molar absorptivity coefficient, for the substance of interest, **b** is the length of the light path, and **C** is the concentration of the absorbing species. Establish an absorbance-concentration calibration curve by preparing at least seven solutions of crystal violet. You will have to select an appropriate range of concentrations based on your observations about crystal violet decolorization. Measure the absorbances of these solutions and determine if there is a straight-line relationship between absorbance and concentration. You will have to conduct an experiment to determine which wavelength of light is best for measuring the absorbance of crystal violet. Be careful handling the crystal violet solution. It can stain skin and clothing.

Your experiments should monitor the reaction until the absorbance of the reaction mixture is 0.20 or less. Make sure not to exceed the total number of data points allowed by your data collection device when designing your experiment.

Report

Your report should include the Beer-Lambert law calibration data and any other pertinent data gathered and/or plotted. The report must be typed and grammatically correct. It may be returned to you for corrections if it is not acceptable.

> **Caution:**
> **Wear your goggles at all times. You are working with strong acids and bases that can damage skin and eyes. Crystal violet is a biological stain.**

How Fast Does the Crystal Violet Decolorize? **Team Contribution Form**

Identify the percent contribution each team member made in this investigation, listing your name first.

Name: Percent Contribution:

_____   _____

_____   _____

_____   _____

_____   _____

Describe your contribution to the completion of the investigation and the writing of the report.

Return this sheet to your lab instructor.

Investigation 31
Why Is the Vinegar Factory Rusting?

Introduction

Rusting of steel is a common problem in vinegar production facilities. One vinegar manufacturer has recently asked your team to help with its rust problem. Beyond knowing that oxygen gas concentration changes as the rust is produced, it knows little about the chemistry or mathematical relationships involved in the problem. The vinegar production company has asked your team to investigate rust production in the presence of acetic acid as a function of changes in gas pressure.

Rate measurements are usually made under conditions such that all of the factors which might affect the rate are held constant, except one, so that the dependence of the rate on each variable is determined separately. Since molecules must come together to react, the rate of a reaction is expected to depend on the concentrations of reactants and possibly other substances. The dependence of the rate on concentrations is represented by a rate law, which is usually of the form:

$$\text{Rate} = k\,[A]^a[B]^b[C]^c$$

where k is the rate constant and the superscripts are called the order with respect to that particular substance.

Goals

As you complete this investigation you will:
1. Develop a method for measuring gas changes to simulate the rusting of steel in vinegar production facilities.
2. Describe the chemistry taking place.
3. Find the mathematical relationship between the pressure change (ΔP) (or volume change, ΔV) and time (t).
4. Report your findings in the usual scientific manner.

Materials

Steel wool
5% acetic acid solution
Volume measuring equipment
Balances
Manometer, pressure sensor, or apparatus to collect gas by water displacement

25 mL syringe with connectors
Tygon tubing
50 mL flask with one-hole stopper fitted with dropper tube
125 mL flask with one-hole stopper fitted with dropper tube
Other supplies by request

Investigation 31

Getting Started

To simulate rusting of steel surfaces in vinegar production factories, you should use steel wool dipped in vinegar, shaking off the excess vinegar. You can monitor rust formation as changes in gas pressure by running the reaction in a chamber attached to a manometer or pressure sensor. Alternatively, you could monitor changes in gas volume. In designing your experiment, you should consider an apparatus that allows you to measure the gas pressure changes of the system.

You should find a mathematical relationship between pressure and time. Which of the following equations best fit your data in a linear fashion (where P is the pressure, k is the rate constant, and t is time)?

$$P = P_{final} + kt$$
$$P - P_{final} = P_{initial} - P_{final} + kt$$
$$\ln(P - P_{final}) = \ln(P_{initial} - P_{final}) - kt$$
$$\frac{1}{(P - P_{final})} = \frac{1}{(P_{initial} - P_{final})} + kt$$

As an alternative to measuring the pressure, you could devise an apparatus that measures changes in volume. If so, you should substitute volume for pressure in the mathematical relationships stated above.

Report

Report your findings in the usual scientific manner. Your report should include a discussion of the appropriate mathematical relationship you discovered. You should also make recommendations to the vinegar company about preventing rust formation in its production facilities.

Caution:
Wear your goggles at all times.

Investigation 31

Why Is the Vinegar Factory Rusting? Team Contribution Form

Identify the percent contribution each team member made in this investigation, listing your name first.

Name: Percent Contribution:

_____   _____

_____   _____

_____   _____

_____   _____

Describe you contribution to the investigation and the writing of the report.

Return this sheet to your lab instructor.

Investigation 32
What Factors Affect the Solubility of Kidney Stones?

Introduction

Many people suffer from kidney stones. Approximately 10% of the residents of developed countries will have a kidney stone sometime during their lifetime. Most kidney stones are made of calcium oxalate, CaC_2O_4, a compound that has only limited solubility in water or body fluids. This compound exists in a state of equilibrium in water:

$$CaC_2O_4(s) \rightleftharpoons Ca^{2+}(aq) + C_2O_4^{2-}(aq)$$

The oxalate ion is the conjugate base of a weak acid, oxalic acid. Oxalic acid dissociates in two steps, each of which is a weak acid–conjugate base equilibrium:

$$H_2C_2O_4(aq) \rightleftharpoons HC_2O_4^-(aq) + H^+(aq)$$
$$HC_2O_4^-(aq) \rightleftharpoons C_2O_4^{2-}(aq) + H^+(aq)$$

Oxalic acid is found in many foods, including broccoli, carrots, parsley, green peppers, Romaine lettuce, spinach, rhubarb, sweet potatoes, baked beans, chocolate, soy sauce, Worchestershire sauce, peanuts, pecans, raspberries, strawberries, beer, and brewed tea.

The National Institutes of Health have given you a grant to investigate the conditions under which calcium oxalate precipitates from solution, They are interested in concentration conditions, acidity conditions, and temperature conditions. They also want you to investigate the potential use of citrate and ethylenediaminetetraacetate (EDTA) ions as a possible means of preventing calcium oxalate precipitation.

Goals

As you complete this investigation you will:
1. Determine the concentration conditions necessary to obtain a calcium oxalate precipitate in water.
2. Determine the effect of acidity and temperature on these conditions.
3. Observe what happens when various substances are added to an equilibrium mixture of calcium oxalate.
4. Describe the chemistry taking place when reagents are added that change the equilibrium system.
5. Explain why the equilibrium position changes when the various variables are changed.

Materials

Investigation 32

Calcium chloride, 0.1 M CaCl$_2$
Sodium oxalate, 0.1 M Na$_2$C$_2$O$_4$
Hydrochloric acid, 1 M HCl
Sodium hydroxide, 1 M NaOH
Sodium citrate, Na$_3$C$_6$H$_5$O$_7$·2H$_2$O
Sodium ethylenediaminetetraacetate, Na$_2$C$_{10}$H$_{14}$N$_2$O$_8$·2H$_2$O (Na$_2$EDTA)
Bunsen burner
Ice
Distilled water
Other supplies by request

Getting Started

Start out by determining the maximum concentrations of calcium ions and oxalate ions that can be in solution before a precipitate starts to form. Then investigate the effect of acidity, temperature, and addition of citrate or EDTA ions on the equilibrium mixture. Do the concentrations needed for precipitation need to be higher or lower when you vary these factors? When designing your experiments, work with only small quantities of materials since your NIH budget is not very large.

Report

Submit a report to the National Institutes of Health describing your results. Indicate the conditions under which calcium oxalate is most likely to precipitate. Your report should include a discussion of pertinent chemical reactions and observations. Make recommendations about the possible medical use of citrate and EDTA ions in preventing kidney stone formation. Also include an explanation of why the various factors affected the equilibrium.

Caution:
Wear your goggles at all times.

What Factors Affect the Solubility of Kidney Stones? **Team Contribution Form**

Identify the percent contribution each team member made in this investigation listing your name first.

Group: Percent Contribution:

_____   _____

_____   _____

_____   _____

_____   _____

Describe your contribution to the completion of the investigation and the writing of the report.

Return this sheet to your lab instructor.

Investigation 33
How Many Chemicals Are in the Vial?

Introduction

Two government agents assigned as sole investigators of bizarre phenomena such as alien sitings, ghost hauntings, and supernatural powers have recently encountered a strange solution. The liquid turns blue when shaken but goes back to colorless when allowed to stand. One of the agents believes that the liquid is extraterrestrial and part of a weird alien plot to contaminate earth's water supply. The other agent is skeptical and believes that there is a rational, scientific explanation for the phenomenon. The skeptical agent has convinced the other one to hire your group to study the mystery liquid. The agents want you to determine how many chemicals might be present in the solution and a possible mechanism(s) for the chemical reaction(s) that may be taking place.

Goals

As you complete this investigation you will:
1. Design experiments that will determine the number of components in the unknown liquid.
2. Determine a mechanism for the reaction(s) that may be taking place.
3. Write a report that provides an explanation for the strange phenomenon.

Materials

Unknown liquid
Ice
Tap water
Other supplies by request

Getting Started

Your instructor will supply you with a vial containing the potential alien solution. You will note that upon shaking, a blue color develops rapidly, then fades away slowly. Your group must deduce as much as possible about the chemistry of this system and postulate a mechanism to explain the observations you make. You might use letters to represent the chemical species you determine might be present even if you don't know their identities. You may do anything you wish with the vial or its contents except add another chemical. However, you can add water if you wish. If you decide to open the vial, be aware that you do not know what is in the empty space above the liquid. It could be an alien gas, or some other type of gas, or the empty space could be a vacuum.

Investigation 33

Report

The report must be typed and grammatically correct. It may be returned to you for corrections if it is not acceptable. Your report should include a discussion of the number of chemicals present in the system. You must also include a possible mechanism for the reaction(s) that might be taking place.

> **Caution:**
> **While working in the laboratory wear your goggles at all times. You do not know the identity of the unknown liquid so use caution when handling or opening the bottle.**

How Many Chemicals Are in the Vial? Team Contribution Form

Identify the percent contribution each team member made in this investigation, listing your name first.

Name: Percent Contribution:

_____   _____

_____   _____

_____   _____

_____   _____

Describe your contribution to the completion of the investigation and the writing of the report.

Return this sheet to your lab instructor.

Investigation 34
What Factors Affect Chemical Equilibrium?

Introduction

Addition of a solution of KNCS to a solution of $Fe(NO_3)_3$ results in a blood-red color. The more KNCS added, the more intense the red color. Aqueous Fe^{3+} and NCS^- combine to form a complex ion in an equilibrium reaction:

$$Fe^{3+}(aq) + NCS^-(aq) \rightleftharpoons FeNCS^{2+}(aq)$$

There are many factors that can change the position of the equilibrium once the reaction appears to have stopped. In this investigation you will identify factors that have an effect on the above reaction.

Goals

As you complete this investigation you will:
1. Observe what happens when various substances are added to an equilibrium mixture.
2. Describe the chemistry taking place when reagents are added that change the equilibrium system.
3. Explain why the equilibrium position changes when the mixture is heated and cooled.
4. Identify patterns about stresses placed on equilibrium systems.
5. Summarize your findings in a written report.

Materials

1 M $Fe(NO_3)_3(aq)$
1 M $NH_4NCS(aq)$
0.10 M $SnCl_2(aq)$
0.1 M $AgNO_3(aq)$
1 M $Na_2HPO_4(aq)$

1 M $NH_3(aq)$
Bunsen burner
Ice
Other supplies by request

Getting Started

You should begin by observing the formation of the iron thiocyanate by placing 8 drops of 1 M $Fe(NO_3)_3$ and 8 drops of 1 M NH_4NCS in 100 mL of water. You should design an experiment that will test the effects on the equilibrium when other reagents are added to the system. Your design must minimize reagent use; that is, when placing the external stresses on the equilibrium you should observe what happens when reagents are added in a drop–by–drop fashion. Upon addition of each reagent identify other chemical reactions that might be occurring. Do any

---

What Factors Affect Chemical Equilibrium?

patterns emerge about the direction in which the equilibrium shifts when particular reagents are added?

In addition to chemical stresses, observe what happens when the equilibrium system is heated (but not to boiling) and cooled. What conclusions can you draw about the exo- or endothermicity of the reaction?

Report

Report your findings in the usual scientific manner. Your report should include a discussion of pertinent chemical reactions and observations.

> **Caution:**
> **Wear your goggles at all times. You are working with a strong base that can damage skin and eyes. Avoid inhaling the ammonia fumes. Silver nitrate can stain skin and clothing.**

Investigation 34

What Factors Affect Chemical Equilibrium? Team Contribution Form

Identify the percent contribution each team member made in this investigation, listing your name first.

Name: Percent Contribution:

_____   _____

_____   _____

_____   _____

_____   _____

Describe your contribution to the investigation and the writing of the report.

Return this sheet to your lab instructor.

Investigation 35
What Is the Formation Constant?

Introduction

Addition of a solution of KNCS to a solution of $Fe(NO_3)_3$ results in a blood-red color. The more KNCS added, the more intense the red color. Aqueous Fe^{3+} and NCS^- combine to form a complex ion in an equilibrium reaction:

$$Fe^{3+}(aq) + NCS^-(aq) \rightleftharpoons FeNCS^{2+}(aq)$$

A theater company has recently discovered this reaction and wants to use it to prepare solutions that look like blood. To help the company control the color, your group needs to know the equilibrium constant for the reaction. This will help the company understand the conditions under which to control the reaction.

The formation of a complex ion (or a coordination compound) can be characterized by its equilibrium constant, which is called the formation constant, K_f. The value of K_f can be calculated using the concentrations of all the aqueous species involved in the equilibrium.

$$K_f = \frac{[FeNCS^{2+}]_{eq}}{[Fe^{3+}][NCS^-]_{eq}}$$

Goals

As you complete this investigation you will:
1. Design a procedure for measuring the concentration of $FeNCS^{2+}$ from the color intensity of a solution.
2. Design and carry out a procedure to measure the equilibrium constant for a complex ion formation reaction.
3. Report your findings in the usual scientific manner.

Materials

0.00200 M KNCS
0.00200 M $Fe(NO_3)_3$
0.200 M $Fe(NO_3)_3$
0.10 M HNO_3
Cuvettes

A device for measuring absorbance
Tissues (preferably lint free)
Volume measuring equipment
Other supplies by request

Investigation 35

Getting Started

Because the FeNCS^{2+} complex ion is a colored species, a device that measures light absorbance may be of some use. The amount of light absorbed by a particular substance dissolved in water depends on the length of the light path through the solution, and on the concentration of the absorbing solute. The absorbance is related to concentration according to the Beer-Lambert Law:

$$A = \varepsilon bC$$

where **A** is the absorbance, ε is a proportionality constant, called the molar absorptivity coefficient, for the substance of interest, **b** is the length of light path, and **C** is the concentration of the absorbing species. Establish an absorbance-concentration calibration curve by preparing at least seven solutions of FeNCS^{2+} complex ion at known concentrations. You will have to select an appropriate range of concentrations based on your observations about the formation of the FeNCS^{2+} complex ion. To ensure that you know the FeNCS^{2+} concentrations, use a constant large excess of Fe^{3+} ion, which will drive the reaction to completion and will prevent the formation of other complex ions, such as Fe(NCS)$_2^+$. An additional problem can be avoided by adding a constant amount of acid to each solution. Addition of acid prevents the formation of colored complex ions such as FeOH^{2+}. Since it is important for the amount of HNO$_3$ to be constant from one solution to another, dilution to a consistent volume must be done with HNO$_3$ and varying amounts of water. Use 0.0020 M Fe(NO$_3$) solution as a blank (100% T) for calibrating the spectrometer or colorimeter. The FeNCS^{2+} concentration at equilibrium is calculated for each solution by assuming all KNCS reacts, which is valid if the Fe^{3+} is in large excess. Measure the absorbances of these solutions and determine if there is a straight-line relationship between absorbance and concentration. You will have to conduct an experiment to determine which wavelength of light is best for measuring the absorbance of FeNCS^{2+}.

Once you have generated a standard curve for the absorbance of the FeNCS^{2+} complex ion, mix five or six combinations of Fe^{3+}, NCS$^-$, and HNO$_3$. Use 5.00 mL of 0.00200 M Fe(NO$_3$)$_3$ for each solution and make the total volume of each solution 10.00 mL with 0.10 M HNO$_3$ and water The amount of NCS$^-$ in each solution should be less than the Fe^{3+} amount. Determine the absorbance of the FeNCS^{2+} complex ion after equilibrium is established in each solution. (Equilibrium is established within a few seconds after the reactants are mixed.)

Since neither Fe^{3+} nor NCS$^-$ is in large excess, the [FeNCS^{2+}]$_{eq}$ cannot be calculated; it must be measured. Once that concentration is known, you should be able to calculate the concentrations of the other ions at equilibrium. Use the equilibrium absorbance and concentration data to determine the value of K$_f$ for this system.

Report

The report must be typed and grammatically correct. It may be returned to you for corrections if it is not acceptable.

> **Caution:**
> **Wear your goggles at all times. You are working with a strong acid that can damage skin and eyes.**

What Is the Formation Constant? Team Contribution Form

Identify the percent contribution each team member made in this investigation, listing your name first.

Name: Percent Contribution:

_____   _____

_____   _____

_____   _____

_____   _____

Describe your contribution to the completion of the investigation and the writing of the report.

Return this sheet to your lab instructor.

Investigation 36
Are Household Items Acidic, Basic, or Neutral?

Introduction

Acid-base behavior is not restricted to chemical laboratories. Acid-base chemistry occurs all around us. For example, a kitchen recipe might call for cream of tartar and baking soda. When the two are mixed in water, a chemical reaction occurs in which carbon dioxide is released through the ingredients. The human body is also a factory for complex acid-base chemistry. The body uses several buffering processes that maintain a constant pH in the various body systems.

Suppose your team of chemical investigators is working for a home safety agency interested in exploring the acidity and basicity of common household items. You have been asked to determine the pH of these items using various techniques.

Goals

As you complete this investigation you will:
1. Determine the pH of various household items.
2. Compare the merits of various techniques for collecting acid-base information.
3. Use the pH data to determine the acid or base content of those common items.
4. Report your findings in a scientific manner.

Materials

Vinegar	Distilled water
Bleach	Water samples from other sources, such as a swimming pool or reservoir
Lemon juice	
Soft drink	pH meter
Liquid dish soap	Litmus paper
Ammonia	Universal indicator
Drain cleaner	Standard pH solutions
Baking soda	Well plate
Tap water	Other supplies by request

Getting Started

In this experiment, you will use litmus paper, universal indicator, and a pH meter to collect acid-base data about the items listed above and any other materials provided by your instructor. A neutral solution has a pH of 7, an acid has a pH below 7, and a base has a pH above 7. (See Appendix C for additional information about measuring pH.) You should consult your textbook

Investigation 36

about information regarding acids and bases so you can adequately define those terms. Use of red and blue litmus paper provides a simple qualitative way of determining if a solution is an acid or a base. Red litmus paper will turn blue in the presence of base and blue litmus paper will turn red in the presence of acid. If needed, consult your text for further information regarding acids and bases.

Universal indicator also provides a simple way of determining the acid or base content of a substance. Typically, a drop of the indicator is added to a small amount of reference solution, yielding a color unique to the pH of the standard.

Finally, the acid-base content of a substance can be determined using a pH meter. Directions for using a pH meter can be found in Appendix C of this manual. .

*Because of the dangerous acid-base properties of some of the provided household items, do **not** mix any together.*

Report

The report must be typed and grammatically correct. It may be returned to you for corrections if it is not acceptable. Your report should include a discussion of the merits of the three techniques for collecting acid-base data

> **Caution:**
> **Wear your goggles at all times. Some of the household substances you are working with are strong acids and bases that can damage skin and eyes. Do not mix any of the household chemicals together.**

Investigation 36

Are Household Items Acidic, Basic, or Neutral? Team Contribution Form

Identify the percent contribution each team member made in this investigation, listing your name first.

Name: Percent Contribution:

_____   _____

_____   _____

_____   _____

_____   _____

Describe your contribution to the completion of the investigation and the writing of the report.

Return this sheet to your lab instructor.

Investigation 37
What Is the pH of Soil?

Introduction

The acidity of soils is very important in the growing of foodstuffs. The pH of the soil is a factor in regulating the availability of plant nutrients and the activity of soil bacteria. In some cases, controlled acidity may help prevent growth of weeds and certain destructive insects. Some plants grow well only in slightly acid soils while excess acidity may prevent good growth with other plants. In alkaline soils (pH > 8), such as are found in desert regions, the amounts of nitrogen, phosphorus, iron, and other nutrients are too low for good plant growth and special procedures such as neutralization by acid treatment must be carried out to make the soil suitable for agriculture. The County Agricultural Extension Service wants you to determine the pH of some soil samples and the effect of various additives on the pH of these soils.

Goals

As you complete this investigation you will:
1. Develop a method for measuring whether a soil is acidic, basic, or neutral.
2. Use your method to determine the pH of some soil samples.
3. Add some fertilizer components to soil samples to make mixtures that are as homogeneous as possible.
4. Measure the effect of the fertilizer components on the pH of the soil samples.

Materials

Soil samples
Soil additives:
    Lime, $Ca(OH)_2$
    Iron(III) sulfate, $Fe_2(SO_4)_3$
    Ammonium sulfate, $(NH_4)_2SO_4$
    Potassium chloride, KCl
    Diammonium hydrogen phosphate, $(NH_4)_2HPO_4$
Red litmus paper (turns blue when wet with a basic solution)
Blue litmus paper (turns red when wet with an acidic solution)
Bromcresol purple solution (yellow if pH < 5.2 and purple if pH > 6.8; green between these pH values)
Distilled water
Glass plate or spot plate
Spatula
Glass rod
Other supplies by request

Investigation 37

Getting Started

Begin by developing a method for determining the acidity of a known soil. Use the two acid-base indicators, litmus and bromcresol purple. A pH meter could also be used if one is available. (See Appendix C for more information about pH.) Your glassware must be very clean or it might cause a change of pH. How can you check this? When you have checked your method on a soil sample of know pH, use it to determine the acidity of the unknown soils. Then, one at a time, add the various soil additives to different soil samples. Do this in such a way as to make the mixture as homogeneous as possible. What quantities will you use? Use your method to determine the effect of these additives on the acidity of the soil.

Report

Submit a report to the Extension Service that describes the method you developed for measuring acidity. Include the results for the various soil samples. Describe the effect on pH of the various soil additives. Also include a description of why these soil additives might be used to assist plant growth.

> **Caution:**
> **Wear your goggles at all times.**

What Is the pH of Soil? **Team Contribution Form**

Identify the percent contribution each team member made in this investigation listing your name first.

Group: Percent Contribution:

_____   _____

_____   _____

_____   _____

_____   _____

Describe your contribution to the completion of the investigation and the writing of the report.

Return this sheet to your lab instructor.

Investigation 38
What Is the Acid Dissociation Constant?

Introduction

A small chemical company with limited personnel resources has recently experienced an identification problem with a production batch of a weak acid. Because of confusion at a production facility, it has 1000 liters of an acid solution but does not know the identity of the acid. The company has asked your savvy research team to investigate the problem. First, it wants you to determine the acid K_a and the molarity so it can identify the production batch. Second, the company wants your interpretation of the graphical relationship between pH and the volume of added NaOH in a titration of its unknown acid.

The strength of an acid can be characterized by its equilibrium constant, which is called the acid dissociation constant, K_a. Using HA to denote a general acid, the equation can be written for the behavior of an acid in water:

$$HA(aq) + H_2O(l) \rightarrow H_3O^+(aq) + A^-(aq)$$

The value of K_a can be calculated using the concentrations of the aqueous species involved in the equilibrium:

$$K_a = \frac{[H_3O^+]_{eq}[A^-]_{eq}}{[HA]_{eq}}$$

Goals

As you complete this investigation you will:
1. Design and carry out an experiment to determine the K_a value for a weak acid.
2. Determine the molarity of a solution of a weak acid.
3. Graphically determine the relationship between pH and added NaOH.
4. Report your findings in the usual scientific manner.

Materials

0.1 M NaOH(aq) (standardized)*
Unknown weak acid solution
Burets and other volumetric measuring equipment

pH meter
Buffer solutions to calibrate pH meter
Other supplies by request

*Obtain exact concentration from reagent bottle.

Investigation 38

Getting Started

One of the most common ways to determine weak acid molarity is by titrating the acid with standardized NaOH solution in the presence of an acid-base indicator. Since no indicators are provided for your use, try to design a procedure that will not require the use of an indicator. You may use the pH meter. Upon approval from your instructor, you might want to begin by doing a quick run-through to develop your skills.

Note that the acid concentration is *approximately* 0.1 *M*. You are to determine the *exact* concentration of the acid to as many significant figures as possible.

You should collect enough data in your experiment to determine the K_a value for the acid. To be confident in your value of K_a, you should run your experiment several times. Your reported K_a would then be an average of several determinations. In addition, you should be able to determine the K_a from the same set of data in more than one way.

Report

Your report should include all pertinent data collected and used to determine the K_a value and the molarity of the unknown acid solution. Also, speculate about the possible identity of the acid. Include and explain relevant graphs with your report. The report must be typed and grammatically correct. It may be returned to you for corrections if it is not acceptable.

**Caution:
Wear your goggles at all times. You are working with acids and bases that can damage skin and eyes.**

What Is the Acid Dissociation Constant? **Team Contribution Form**

Identify the percent contribution each team member made in this investigation, listing your name first.

Name: Percent Contribution:

_____   _____

_____   _____

_____   _____

_____   _____

Describe your contribution to the completion of the investigation and the writing of the report.

Return this sheet to your lab instructor.

Investigation 39
What Is the Solubility Product?

Introduction

The water in swimming pools generally contains a considerable amount of dissolved calcium ions. Pool water is purified by the addition of some chlorinating agent, often calcium hypochlorite. In addition, calcium ions can arise from the plaster that lines the pool. Plaster is a hydrated form of calcium sulfate. A swimming pool company has recently contacted your group to address a complaint from some of its customers. The customers claim that the plaster in their one-year-old pools is dissolving. The company wants to know how much plaster would dissolve before the pool water became a saturated solution of calcium sulfate, if a pool were filled initially with softened water containing no calcium ions.

The solubility of a compound can be characterized by the equilibrium constant that corresponds to an equation representing the dissolution of the compound. The equilibrium constant is called the *solubility product*, K_{sp}. The dissociation in water of calcium sulfate can be written as:

$$CaSO_4(s) \rightleftharpoons Ca^{2+}(aq) + SO_4^{2-}(aq)$$

The value of K_{sp} can be calculated using the concentrations of the aqueous species involved in the equilibrium.

$$K_{sp} = [Ca^{2+}][SO_4^{2-}]$$

Goals

As you complete this investigation you will:
1. Develop a method for measuring the amount of calcium sulfate dissolved in a saturated solution and the value of K_{sp} for calcium sulfate.
2. Describe a typical swimming pool and calculate the mass of calcium sulfate that would be removed from the plaster pool lining to produce a saturated solution.
3. Report your findings in the usual scientific manner.

Materials

$CaSO_4(s)$
A device for measuring conductivity
Common laboratory equipment
Other supplies by request

Investigation 39

Getting Started

In this investigation you will design and carry out an experiment to determine the solubility product constant for CaSO$_4$. Some points to consider include how to ensure that the solution you prepare is a saturated solution and how to measure the amount of dissolved calcium sulfate.

A possible approach involves the measurement of conductivity. The conductivity of a solution is directly proportional to the molarity of the dissolved ions. If you assume all dissolved CaSO$_4$ exists as individual Ca^{2+} and SO$_4^{2-}$ ions, then the relationship between molarity (solubility) and K$_{sp}$ is not complicated.

Report

The report must be typed and grammatically correct. It may be returned to you for corrections if it is not acceptable. You should report your K$_{sp}$ and compare its value to that found in the literature. Also include the information requested by the swimming pool company.

**Caution:
Wear your goggles at all times.**

What Is the Solubility Product? **Team Contribution Form**

Identify the percent contribution each team member made in this investigation, listing your name first.

Name: Percent Contribution:

_____   _____

_____   _____

_____   _____

_____   _____

Describe your contribution to the completion of the investigation and the writing of the report.

Return this sheet to your lab instructor.

Investigation 40
What Are Some Chemical Properties of Cream of Tartar?

Introduction

Several simple chemical compounds have very practical purposes in a kitchen. Potassium hydrogen tartrate, $KHC_4H_4O_6$, is the ingredient in a compound commonly known as cream of tartar. Sodium bicarbonate is found in baking soda and baking powder. When the acidic hydrogen tartrate anion reacts with a bicarbonate salt during baking, a leavening process occurs in which carbon dioxide is produced. The hydrogen tartrate won't react until it's dissolved in the batter. The action of cream of tartar is most obvious in recipes for items like angel food cake.

A group of entrepreneurs wishing to cash in on the boom in the home products market has hired your team to study the behavior of potassium hydrogen tartrate (abbreviated KHT). The entrepreneurs want your team to investigate the room temperature solubility of potassium hydrogen tartrate in water. They want to use the information to develop a "baking" process that doesn't require heat. They hope to capture a market for "fresh-baked" products that can be prepared without an oven. The entrepreneurs want you to find the K_{sp} value for KHT at room temperature. In addition, to understand the acid-base properties of cream of tartar, they want you to determine the K_a value for the hydrogen tartrate anion. They are willing to pay your group a fair rate for its services. If you are successful, then they may consider your group for future work.

Goals

As you complete this investigation, you will:
1. Develop a procedure to determine the K_{sp} of potassium hydrogen tartrate.
2. Use the data, or develop another procedure, to determine the K_a value of the hydrogen tartrate anion.
3. Prepare a report and a bill for laboratory services.

Materials

$KHC_4H_4O_6(s)$
0.05 M NaOH(aq)[*]
pH meter
Standard buffer solutions
Other supplies by request

Magnetic stirrer (optional)
Common laboratory equipment
Filtration apparatus
Buret

   [*]*Obtain exact concentration from reagent bottle.*

Investigation 40

Getting Started

The solubility of a compound can be characterized by the equilibrium constant that corresponds to an equation representing the dissolution of the compound. The equilibrium constant is called the *solubility product*, K_{sp}. The dissociation in water of potassium hydrogen tartrate can be written as:

$$KHC_4H_4O_6(s) \rightleftharpoons K^+(aq) + HC_4H_4O_6^-(aq)$$

The value of K_{sp} can be calculated using the concentrations of the aqueous species involved in the equilibrium.

$$K_{sp} = [K^+][HC_4H_4O_6^-]$$

In contrast, the strength of an acid can be characterized by its equilibrium constant, which is called the *acid dissociation constant*, K_a. Using HA to denote a general acid, the equation can be written for the behavior of an acid in water:

$$HA(aq) + H_2O(l) \rightleftharpoons H_3O^+(aq) + A^-(aq)$$

The value of K_a can be calculated using the concentrations of the aqueous species involved in the equilibrium:

$$K_a = \frac{[H_3O^+]_{eq}[A^-]_{eq}}{[HA]_{eq}}$$

The hydrogen tartrate anion is a weak acid and will react with hydroxide. A standard hydroxide solution could be used to determine the concentration of hydrogen tartrate in a solution by titration. When you prepare your potassium hydrogen tartrate solution, consider how you will ensure that it is saturated. To be confident in your experimental values of K_{sp} and K_a, you should run your experiment several times.

Report

The report must be typed and grammatically correct. It may be returned to you for corrections if it is not acceptable. Remember to prepare a bill for your client. The entrepreneurs are willing to pay $100 per hour for your services in addition to chemical and supply costs, if your report is satisfactory.

> **Caution:**
> Wear your goggles at all times. You are working with acids and bases that can damage skin and eyes.

Investigation 40

What Are Some Chemical Properties of Cream of Tartar? Team Contribution Form

Identify the percent contribution each team member made in this investigation, listing your name first.

Name: Percent Contribution:

_____   _____

_____   _____

_____   _____

_____   _____

Describe your contribution to the completion of the investigation and the writing of the report.

Return this sheet to your lab instructor.

Investigation 41
What Are the Metals?

Introduction

Metal ions and metals exhibit characteristic values of reduction potentials. These values are of interest because they can be used to predict possible oxidation-reduction reactions between metals and their ions. This information can also be used to design batteries made up of two metals and their ions. A company which specializes in battery production wants your group to investigate five unknown metals for their potential to provide voltage. The company wants you to use the data you collect to predict the identities of the metals and to describe the best battery that can be made from a combination of two metals and their corresponding electrolyte solutions.

Goals

As you complete this investigation you will:
1. Develop a procedure for measuring the voltage developed by galvanic cells constructed from combinations of five metals and solutions of their salts.
2. Determine the reduction potentials and identities of the five metals.
3. Outline a battery design using two of the five metals and their corresponding electrolyte solutions that will give the largest voltage possible with these materials.
4. Report your findings in the usual scientific manner.

Materials

5 metals identified only as M_1 through M_5
5 metal ion solutions (same molarities) identified only as M_1 through M_5, matching the metals
1.0 M NaNO$_3$(*aq*)
Glass dishes, well plates, or small beakers
Filter paper, yarn, or cotton string
A device for measuring voltage

Getting Started

Note that one of the metals, M_1, is distinct in its appearance from the other metals, which should allow you to identify it. This metal should make a useful reference in identifying the other metals. Be sure to clean the surface of the metals with steel wool before use.

You might begin by placing a piece of filter paper in a glass dish. Then place several drops of the ion solution that corresponds to the metal labeled M_1 on the filter paper. Put a piece of M_1 in the "puddle." You have just created a half-cell. Use another metal and its solution to make a second

Investigation 41

half-cell. The two half-cells must be connected in some way to complete the circuit. There are several items available that might serve this purpose. Are there any materials available that could be used as a salt bridge? Consult Appendix F for instructions about measuring voltage with your equipment. How will you use the voltage data to determine the identity of the unknown metal? A table of standard reduction potentials in your textbook might be of some use. Repeat your procedure for the other metals. Will you use the M_1 half-cell as the reference for all your tests? Once you've identified other metals, their half-cells could serve as additional references.

If you are unable to use your voltage data to conclusively identify the metals, consider other experiments you could run. What other properties of the metals might help you distinguish one from the other? Your instructor must approve additional experiments you plan to carry out.

Report

The report must be typed and grammatically correct. It may be returned to you for corrections if it is not acceptable. Be sure to include your description of the best battery. What are the practical limitations of this battery?

> **Caution:**
> **Wear your goggles at all times. Do not touch the solutions.**

Investigation 41

What Are the Metals? Team Contribution Form

Identify the percent contribution each team member made in this investigation, listing your name first.

Name: Percent Contribution:

_____   _____

_____   _____

_____   _____

_____   _____

Describe your contribution to the completion of this investigation and the writing of the report.

Return this sheet to your lab instructor.

Investigation 42
How Can a Battery Be Made from Coins?

Introduction

In 1800 Alessandro Volta created the first battery, which consisted of alternating layers of zinc, paper soaked in salt water, and silver. This battery has come to be known as a voltaic pile. Your group can create your own pile by using coins and paper towels. Suppose your group was stranded on a boat somewhere in the ocean and you have to power up your GPS (global positioning system). You have no replacement batteries but you have a pocket full of coins. How much voltage can you generate from these coins to make a battery?

Goals

As you complete this investigation you will:
1. Develop a procedure for measuring the voltage of a pile of coins.
2. Vary the coin composition and arrangement.
3. Report your findings in the usual scientific manner.

Materials

Coins (bring your own)
NaCl(*s*)
Filter paper or paper toweling

A device for measuring voltage
Acetic acid and/or steel wool to clean the coins

Getting Started

You might begin by constructing a voltaic pile consisting of pennies and dimes. Try starting with a saturated solution of sodium chloride for soaking the paper. Alternate layers of pennies, paper soaked in salt water, and dimes. The coins should be cleaned with some acetic acid and/or steel wool before you begin. Measure the voltage of the pile. Does the voltage vary with pile height? If so, does it vary consistently with additional layers? From your saturated NaCl solution, prepare other concentrations. How does the voltage vary with different concentrations of salt water? How does the voltage vary with different types of coins? Consult Appendix F for instructions about measuring voltage with your equipment.

Investigation 42

Report

The report must be typed and grammatically correct. It may be returned to you for corrections if it is not acceptable. Include a thorough discussion of your observations, answering the questions posed in this investigation.

**Caution:
Wear your goggles at all times.**

Investigation 42

How Can a Battery Be Made from Coins? **Team Contribution Form**

Identify the percent contribution each team member made in this investigation, listing your name first.

Name: Percent Contribution:

_____   _____

_____   _____

_____   _____

_____   _____

Describe your contribution to the completion of this investigation and the writing of the report.

Return this sheet to your lab instructor.

Investigation 43
What Is the Complex Ion?

Introduction

Ions formed between transition metal ions and molecular or ionic species called *ligands* do not conform to the usual composition of chemical compounds (generally determined by a condition of electrical neutrality). These ions are called *complex ions*. Many of these ions can lose the ligands readily, so it may not be possible to isolate the compounds from solution to determine the composition. Instead, a procedure must be developed that will measure the composition of the ion in solution. Fortunately, most of these complex ions are colored, while their components are often colorless. Thus, the absorbance of solutions at an appropriate wavelength can often be used to measure the amount of a complex ion in solution.

In this investigation you will design and carry out an experiment to determine the formula for the complex which forms between Fe^{2+} and ortho-phenanthroline (o-phen, a ligand); that is, you will find x and y for the formula:

$$[Fe_x(\text{o-phen})_y]^{2+}$$

Goals

As you complete this investigation you will:
1. Develop a procedure for determining the formula of a complex ion.
2. Determine the formula of the complex ion formed from iron(II) ion and o-phenanthroline.
3. Report your findings in the usual scientific manner.

Materials

0.0050 M Fe(NH$_4$)$_2$(SO$_4$)$_2$(*aq*)* (ferrous ammonium sulfate)
0.0050 M ortho-phenanthroline solution*
Volume measuring equipment
Device for measuring absorbance
1 cm cuvettes
Tissues (preferably lint free)
   *Obtain exact concentration from reagent bottle.

Getting Started

The complex ion you will study is a colored species so use of an absorbance measuring device may be appropriate. You will have to determine the wavelength at which the complex best absorbs light.

Investigation 43

Most complex ions have a simple formula involving one metal ion and 1–6 ligands. Keep this in mind when designing your experiment because you will not be able to use a standard curve technique as you may have done in previous experiments. Instead you should vary the proportion of metal ion to ligand in a way that will allow you to determine the correct formula for the complex ion.

Report

Your report should include all pertinent data collected and/or plotted to obtain the formula for the complex. You should also use your results to predict the structure of the complex ion. The report must be typed and grammatically correct. It may be returned to you for corrections if it is not acceptable.

**Caution:
Wear your goggles at all times.**

What Is the Complex Ion? Team Contribution Form

Identify the percent contribution each team member made in this investigation, listing your name first.

Name: Percent Contribution:

_____   _____

_____   _____

_____   _____

_____   _____

Describe your contribution to the completion of this investigation and the writing of the report.

Return this sheet to your lab instructor.

Investigation 44
What Formulation Makes the Best Toy?

Introduction

A toy manufacturer wishes to break into a specialized toy market that depends on various properties of polymers. Using a combination of glue and sodium borate they believe that a slime-type of polymer using polyvinyl acetate can be developed. They hope this unique product can be packaged and sold in toy stores. The company wants your group to produce several variations of glue/sodium borate combinations and compare their properties. The group that can produce a ball with the best bounce will win the company's contract.

Goals

As you complete this investigation you will:
1. Develop formations of polymers containing polyvinyl acetate.
2. Compare viscosity properties of the different formulations.
3. Develop a combination of raw materials that produce a final product that will bounce.
4. Report findings to the toy manufacturer.

Materials

White glue (polyvinyl acetate)
Sodium borate
Balances
Graduated cylinder, burets, pipets, or other volume measuring devices
Hot plates
Other supplies by request

Getting Started

In this investigation you will prepare several different products that contain various amounts of polyvinyl acetate, water and sodium borate. By varying the ratios of these starting materials, properties such as viscosity might change. Because of the complexity of the reaction taking place, you cannot merely add water and solid sodium borate to glue. Instead, some glue has to be mixed with water, the sodium borate has to be dissolved in water, and the two solutions combined. You should begin your first formulation using a 1:1:1:1 mass ratio of each component – 10 gram quantities should suffice. Start by dissolving 10 g glue in 10 g of distilled water. (Assume a density of 1 g/mL for distilled water.) Also dissolve 10 g of sodium borate in 10 g of distilled water. The two solutions can be mixed by adding the glue mixture to the sodium borate solution. You may have to work the mixture with your hands to thoroughly mix the components.

Investigation 44

When you prepare variations of the 1:1:1:1 ratio of glue/water/sodium borate/water, you should keep the total mass constant. Develop a technique for comparing the viscosity and bouncability of each formulation.

Report

The report must be typed and grammatically correct. It may be returned to you for corrections if it is not acceptable. Include your recommendations to the toy manufacturer.

> **Caution:**
> **Wear your goggles at all times. Sodium borate is lightly toxic if inhaled or ingested. Some people are allergic to soaps containing sodium borate so wash your hands frequently during this investigation.**

What Formulation Makes the Best Toy? **Team Contribution Form**

Identify the percent contribution each team member made in this investigation, listing your name first.

Name: Percent Contribution:

_____   _____

_____   _____

_____   _____

_____   _____

Describe your contribution to the completion of the investigation and the writing of the report.

Return this sheet to your lab instructor.

Investigation 45
How Are Anions Identified?

Introduction

Chemists are often concerned with the identification of species present in a substance – qualitative analysis. They use a variety of techniques involving chemical or instrumental analyses. In this investigation your team will propose and carry out a chemical analysis scheme for identification of anions present in aqueous solution. A company that manufactures on-site test kits has contracted for your team's services to provide background chemical information on its anion identification project. The company wants your thorough analysis of the chemistry involved in separating and confirming bromide, carbonate, chloride, iodide, phosphate, sulfate, and sulfite ions using reagents readily available to it for its kits. A chemist previously employed by the company developed the attached flow chart but left the firm before completing the description of the chemistry. To test your procedures, the company has sent a known mixture of the anions (each at 0.1 M) and an unknown mixture. The company wants separate reports from each team member to assess the reliability of your testing procedures.

Goals

As you complete this investigation you will:
1. Develop procedures for identification and confirmation of known anions present in solution.
2. Use these procedures for identification of anions in an unknown mixture.
3. Report your findings in the usual scientific fashion.

Materials

Known mixture of anions
Individual anion solutions
6 M HNO$_3$(aq)
0.1 M AgNO$_3$(aq)
6 M NH$_3$(aq)
15 M NH$_3$(aq)

0.1 M Ba(NO$_3$)$_3$(aq)
0.5 M (NH$_4$)$_2$MoO$_4$(aq)
0.02 M KMnO$_4$(aq)
Saturated Ba(OH)$_2$
Centrifuge
Other supplies by request

Getting Started

Use the flow chart as a guide for developing your procedures. You'll notice that the steps listed in the flow chart divide the ions into manageable groups. These groups are, of course, based on the chemical properties of the species involved. You should begin by writing equations for the reactions in the qualitative analysis scheme. The steps listed are destructive to your sample so you must be conscious of sample size issues. For a discussion of semimicro analysis techniques and anion chemistry, you could begin by consulting the information available at the Internet site,

Investigation 45

http://www.public.asu/~jpbirk/, or other appropriate references. After developing your analysis scheme you should test it on the known solution first. Following analysis of the known, a unique unknown will be issued to each member of your group.

Report

Each student must submit a separate report. Your discussion must include a thorough explanation of the chemical reactions. You must also write equations for the reactions as noted in the above flow chart.

> **Caution:**
> **While working in the laboratory wear your goggles at all times. You are working with strong acids and bases that can cause damage to skin and eyes. Also, some of the chemicals you are working with are toxic.**

Analysis of Anions Flow Chart

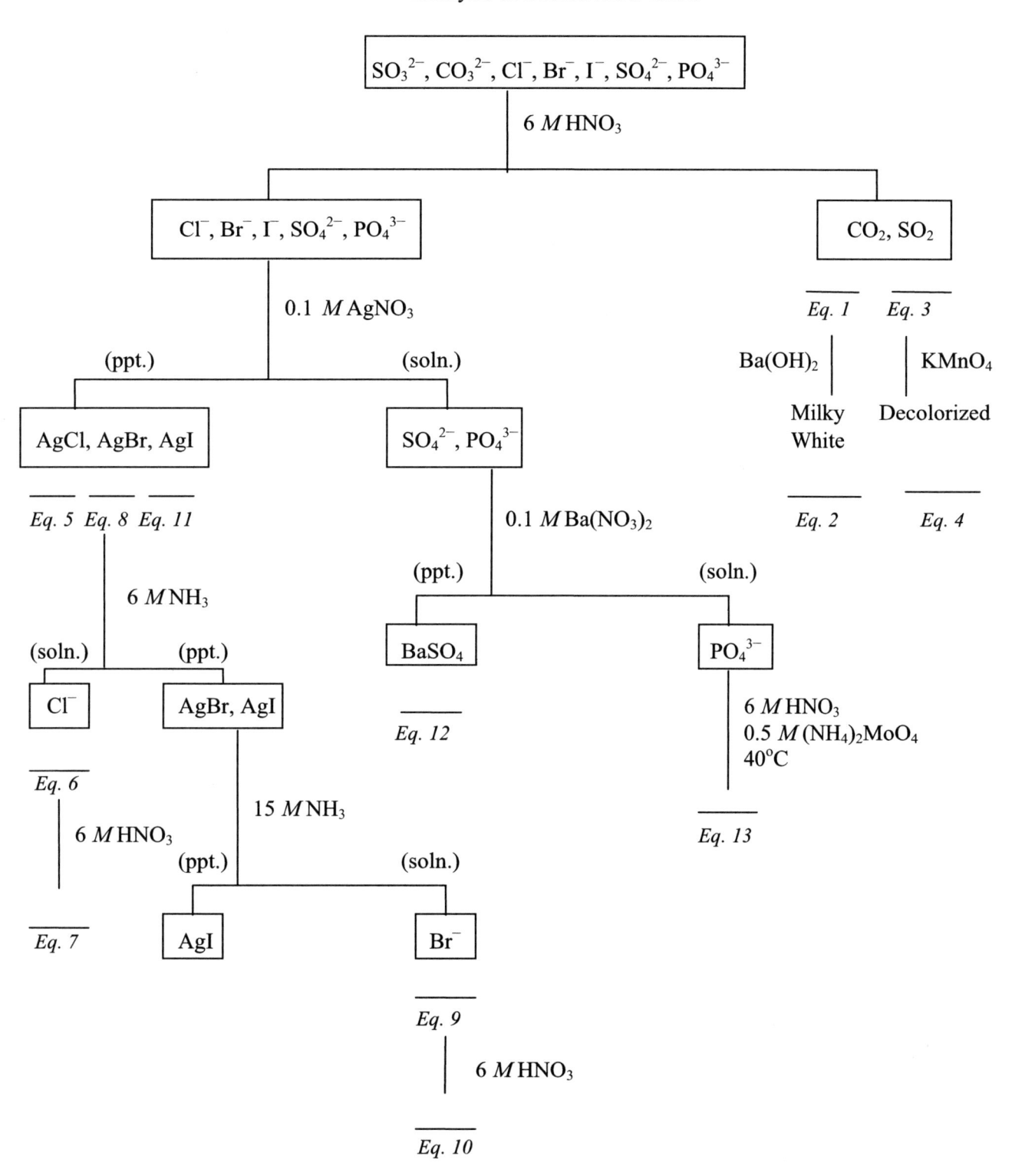

$SO_3^{2-}, CO_3^{2-}, Cl^-, Br^-, I^-, SO_4^{2-}, PO_4^{3-}$

↓ 6 M HNO_3

Branch A: $Cl^-, Br^-, I^-, SO_4^{2-}, PO_4^{3-}$

Branch B: CO_2, SO_2
- $Ba(OH)_2$ → Milky White — Eq. 1, Eq. 2
- $KMnO_4$ → Decolorized — Eq. 3, Eq. 4

Branch A treated with 0.1 M $AgNO_3$:

- (ppt.) AgCl, AgBr, AgI — Eq. 5, Eq. 8, Eq. 11
- (soln.) SO_4^{2-}, PO_4^{3-}

AgCl, AgBr, AgI + 6 M NH_3:
- (soln.) Cl^- — Eq. 6; then 6 M HNO_3 — Eq. 7
- (ppt.) AgBr, AgI

AgBr, AgI + 15 M NH_3:
- (ppt.) AgI
- (soln.) Br^- — Eq. 9; then 6 M HNO_3 — Eq. 10

SO_4^{2-}, PO_4^{3-} + 0.1 M $Ba(NO_3)_2$:
- (ppt.) $BaSO_4$ — Eq. 12
- (soln.) PO_4^{3-} — 6 M HNO_3, 0.5 M $(NH_4)_2MoO_4$, 40°C — Eq. 13

How Are Anions Identified?

How Are Anions Identified? Team Contribution Form

Identify the percent contribution each team member made in this investigation, listing your name first.

Name: Percent Contribution:

_____   _____

_____   _____

_____   _____

_____   _____

Describe your contribution to the investigation.

Return this sheet to your lab instructor.

Investigation 46
How Are Cations Identified?

Introduction

In this investigation your team will propose and carry out a chemical analysis scheme for identification of cations present in aqueous solution. A company that manufactures on-site test kits has contracted for your team's services to provide background chemical information on its cation identification project. The company wants your thorough analysis of the chemistry involved in separating and confirming selected cations assigned to your group from the following list:

Ag^+	Co^{2+}	Mn^{2+}
Al^{3+}	Cr^{3+}	Ni^{2+}
Ba^{2+}	Cu^{2+}	Pb^{2+}
Ca^{2+}	Fe^{3+}	Zn^{2+}
Cd^{2+}	Hg_2^{2+}	

You can only use the reagents listed below. The company does not employ a chemist who could devise a flow chart for the analysis so your team will have to develop one. In addition to your flow chart the company wants a description of the chemistry involved in your scheme. To test your procedures the company has sent samples of the individual cations (at 0.2 *M*), a list of potential cation mixtures, and an unknown mixture.

Goals

As you complete this investigation you will:
1. Develop procedures for identification and confirmation of known cations present in a solution.
2. Develop a flow chart that schematically represents the procedures you carry out.
3. Use these procedures for identification of cations in an unknown mixture.

Materials

Nitrate salts of individual cations (each at 0.2 *M*)
6 *M* $NH_3(aq)$
6 *M* $NaOH(aq)$
6 *M* $HCl(aq)$
6 *M* $HNO_3(aq)$
3 *M* $H_2SO_4(aq)$
Centrifuge
Other supplies by request

Investigation 46

Getting Started

A known mixture of four cations will be assigned to your group. Before beginning the lab work, one of your team members might obtain chemical information about the behavior of the cations you will study. For a discussion of the chemistry of various cations, you could begin by consulting the information available at the Internet site, *http://www.public.asu/~jpbirk/*, or other appropriate references. Solubility rules will serve as the basis for many of the reactions you will carry out, so you should also obtain appropriate solubility information. For the cations your group has been assigned, you should carry out reactions to experimentally confirm the information in your literature search. After studying the chemistry of the individual cations, you should develop a plan of action for identifying and confirming the cations. To achieve greater control of the reactions, you may want to dilute your cation solutions — you might use your initial observation data as a guide in making the decision to dilute.

To organize your findings, you must develop a flow chart to schematically represent your procedures for separating and confirming the cations in your assigned known mixture. To avoid generating excess chemical waste, prepare your known mixture with <1 mL samples of the cations. Since you will be carrying out reactions that are destructive to your sample, you must be conscious of sample size issues — you have a limited amount of each cation solution available so you must use it sparingly. For a discussion of semimicro analysis techniques, consult the information available at the Internet site listed above or other appropriate references. Someone in your group should assume the duties of finding background chemical and procedural information for your team.

After developing your analysis scheme, you should test it on a known mixture first. As directed by your instructor, your group could prepare samples for each other to test your flow chart. Following analysis of the known mixture, a unique unknown will be issued to each member of your group, consisting of cations for which you have developed a scheme. You might consider running a known mixture of your assigned cations alongside the unknown.

Report

Each student must submit a separate report. Your report must include the flow chart followed and a discussion of the chemical reactions that took place in identifying your unknown cations.

Caution:
While working in the laboratory wear your goggles at all times. You are working with strong acids and bases that can cause damage to skin and eyes. Also, some of the chemicals you are working with are toxic.

How Are Cations Identified? Team Contribution Form

Identify the percent contribution each team member made in this investigation, listing your name first.

Name: Percent Contribution:

_____   _____

_____   _____

_____   _____

_____   _____

Describe your contribution to the investigation.

Return this sheet to your lab instructor.

Investigation 47
How Are More Cations Identified?

Introduction

In this investigation your team will propose and carry out a chemical analysis scheme for identification of a complex mixture of cations in aqueous solution. A company that manufactures on-site test kits has contracted for your team's services to provide background chemical information on its cation identification project. The company wants your thorough analysis of the chemistry involved in separating and confirming eight of the possible cations listed below:

Ag^+	Ba^{2+}	Cd^{2+}	Cu^{2+}	Hg^{2+}	Ni^{2+}	Sr^{2+}
Al^{3+}	Bi^{3+}	Co^{2+}	Fe^{3+}	Mg^{2+}	Pb^{2+}	Zn^{2+}
As^{3+}	Ca^{2+}	Cr^{3+}	Hg_2^{2+}	Mn^{2+}	Sn^{4+}	

Given the primary reagents listed below and secondary reagents for confirmatory tests, you should develop procedures for separation and identification of the known cations you are assigned. The company has sent samples of the individual cations (at 0.2 M), a list of potential cation mixtures (to be assigned by your lab supervisor), and an unknown mixture containing three to five of the cations assigned to your group. Your group will have to develop a flow chart. In addition to the flow chart, you must write a description of the chemistry involved in your scheme.

Goals

As you complete this investigation you will:
1. Develop procedures for identification and confirmation of eight known cations present in solution.
2. Develop a flow chart that schematically represents the procedures you carry out.
3. Use these procedures to identify cations in an unknown mixture.

Materials

Nitrate salts of individual cations (each at 0.2 M)
6 M NH$_3$(*aq*)
6 M NaOH(*aq*)
6 M HCl(*aq*)
6 M HNO$_3$(*aq*)
3 M H$_2$SO$_4$(*aq*)

1 M thioacetamide (only on request)
Secondary reagents as needed
Centrifuge
Other supplies by request

Investigation 47

Getting Started

A known mixture of eight cations will be assigned to your group. Before beginning the lab work, one of your team members should obtain chemical information about the behavior of the cations you will study. This information is available from several resources, including the Internet site *http://www.public.asu/~jpbirk/*. Solubility rules will serve as the basis for the reactions you will carry out, so you should also obtain appropriate solubility information. When you research the chemistry of your assigned cations, you may notice that many of them precipitate as sulfides with thioacetamide serving as a source of sulfide ion. However, there are particular hazards associated with thioacetamide so you should avoid its use in development of your scheme. You must consider all other possibilities before approaching your instructor about thioacetamide use.

You should carry out preliminary tests on your group's individual cations to experimentally confirm the information in your literature search. After experimentally studying the chemistry of the individual cations, you should develop a plan of action for identifying and confirming the cations. To achieve greater control of the reactions, you may want to dilute your cation solutions — you might use your initial observation data as a guide in making the decision to dilute.

Based on your preliminary testing, you should develop a flow chart to schematically represent your procedures for separating and confirming the cations in your assigned known mixture. To avoid generating excess chemical waste, prepare your known mixture with 0.5 mL samples of the cations. Since you will be carrying out reactions that are destructive to your sample, you must be conscious of sample size issues — you have a limited amount of sample available so you must use it sparingly. For discussion of semimicro analysis techniques, consult the information available at the Internet site listed above or other appropriate references. Someone in your group should assume the duties of finding background chemical and procedural information.

After developing your analysis scheme and flow chart, you should test it on a known solution. As directed by your instructor, your group could prepare samples for each other to test your flow chart. Following this analysis, a unique unknown will be issued to each member of your group. You might consider running a known mixture of your assigned cations alongside the unknown.

Report

Each student must submit a separate report. Your report must include the flow chart followed and a discussion of the chemical reactions that took place in identifying your unknown cations.

> **Caution:**
> **While working in the laboratory wear your goggles at all times. You are working with strong acids and bases that can cause damage to skin and eyes. Also, some of the chemicals you are working with are toxic.**

How Are More Cations Identified? **Team Contribution Form**

Identify the percent contribution each team member made in this investigation, listing your name first.

Name: Percent Contribution:

_____   _____

_____   _____

_____   _____

_____   _____

Describe your contribution to the investigation.

Return this sheet to your lab instructor.

Investigation 48
How Are Ionic Solids Identified?

Introduction

In this investigation you will identify the ions present in an unknown solid mixture. You will rely on information from your previous qualitative analysis investigations. A mining company has located an area containing large deposits of a curious solid mixture. They are able to determine that the mixture consists of two ionic compounds, but they need your help to identify the component cations and anions. It is assumed that you have worked with these ions in your previous investigations.

Goals

As you complete this investigation, you will:
1. Make detailed observations of properties of the solid mixture.
2. Carry out any necessary separations or dissolutions to prepare the mixture for analysis
3. Use an anion flow-chart to assist in identifying the two anions in the unknown mixture.
4. Use portions of your cation flow-charts to assist in identifying the two cations in the unknown mixture.

Materials

Unknown solid mixture
Known anion solutions
Nitrate salts of individual cations (each at 0.2 M)
Concentrated $H_2SO_4(aq)$, $HCl(aq)$, $HNO_3(aq)$
Saturated $Ba(OH)_2$
Saturated Na_2CO_3
Secondary reagents as needed
Centrifuge
Other supplies by request

0.1 M $AgNO_3(aq)$
0.1 M $Ba(NO_3)_3(aq)$
0.5 M $(NH_4)_2MoO_4(aq)$
0.02 M $KMnO_4(aq)$
6 M $NH_3(aq)$
15 M $NH_3(aq)$
6 M $NaOH(aq)$
6 M $HCl(aq)$
6 M $HNO_3(aq)$
3 M $H_2SO_4(aq)$

Getting Started

You will compile a list of clues to help you with your ion identifications. Simple observations like color and solubility may be very important. Published solubility rules will reveal which cation/anion combinations are water-soluble and which combinations are not. The known solutions will reveal which ions are colored in solution.

You will eventually need to get the compounds into an aqueous solution. If one or both ionic

Investigation 48

compounds are insoluble in water, then try to dissolve the solid in increasing strengths of HCl or HNO$_3$. Remember that HCl will introduce Cl$^-$ ions. Also, adding acid may decompose some anions. An acid solution will usually work well for cation analyses.

Another way to get the anions into solution is to boil the solid with saturated Na$_2$CO$_3$ (aq). This will precipitate the cations as their carbonate salts, and the anions will be in solution. This procedure, of course, introduces CO$_3^{2-}$.

Some ionic solids will react with concentrated sulfuric acid. See the table below for a summary of reactions with H$_2$SO$_4$:

Solid Contains	Observation when concentrated H$_2$SO$_4$ is added
Cl$^-$	Effervescence. HCl (g) is evolved. Sharp odor. Turns moist litmus red.
Br$^-$	Effervescence. Brown HBr (g) is evolved. Sharp odor. Turns moist litmus red.
I$^-$	Effervescence. Violet I$_2$ (g) is evolved. Solid turns brown. Smells of H$_2$S
CO$_3^{2-}$	Effervescence. No color or odor.
SO$_3^{2-}$	Effervescence. Colorless gas with a sharp, choking odor.

Keep in mind that observations can be masked when combined with other observations. For example, a yellow colored ion may be present with a blue colored ion, yielding a green solution. A solid may react with concentrated H$_2$SO$_4$ to produce a violet gas, indicating I$^-$, but not necessarily excluding Cl$^-$, Br$^-$, CO$_3^{2-}$, SO$_3^{2-}$, SO$_4^{2-}$, or PO$_4^{3-}$. Useful information is posted at the web site: *http://www.public.asu/~jpbirk/*.

Report

Your report should include all observations pertinent to the identification of the ions. The report should also include any flow-charts used and a description of the chemical reactions involved.

Caution:
While working in the laboratory wear your goggles at all times. You are working with strong acids and bases that can cause damage to skin and eyes. Also, some of the chemicals you are working with are toxic.

Investigation 48

How Are Ionic Solids Identified? Team Contribution Form

Identify the percent contribution each team member made in this investigation, listing your name first.

Name: Percent Contribution:

_____   _____

_____   _____

_____   _____

_____   _____

Describe your contribution to the investigation.

Return this sheet to your lab instructor.

Presentations and Poster Sessions

At the end of the semester all students are expected to present the findings of an investigation in two forms: an oral presentation to your fellow class members and on a large poster board. At the beginning of the lab period your group will do a ten–minute presentation that summarizes your findings. The exposition need not be of a formal scientific type — feel free to pursue a more creative avenue for your presentation.

Scientists often present their work at scientific meetings in the form of poster sessions. The poster typically summarizes the investigation in words, graphs, tables, and figures. Your instructor might direct you to examples scattered around your campus. The poster session will be held immediately after your class's presentations. You are expected to wander by your fellow students' posters in your class and in other sections. In 15 minute shifts, your group members will take turns standing by your poster to field questions from people who are interested in your work. Someone should be near your poster at all times. Expect your fellow students, your instructor, and other guests to ask you questions about your experimental methods and results. You should also ask questions of the other groups as you browse their work.

Your instructor will grade your presentation and poster according to the following criteria:

- Your group's ability to convey an understanding of the investigation during the presentation.
- Your group's presentation skills.
- The creativity of your presentation.
- The scientific accuracy of your findings.
- Your poster's ability to convey an adequate representation of the experiment.
- Proper use of tables, diagrams, and graphs where appropriate.
- The organization and clarity of your poster.
- Your ability to field questions from people browsing through the posters.

Your poster must be visually appealing. Some students make the mistake of merely pasting their lab reports onto a piece of poster board. Also, your reader must be able to see the text from some distance away from the poster.

Presentation/Poster Session Team Contribution Report Form

Identify the percent contribution each team member made to the *presentation*, listing your name first.

Name: Percent Contribution:

_____   _____

_____   _____

_____   _____

_____   _____

Describe your contribution to the *presentation*.

Identify the percent contribution each team member made to the preparation of the *poster*, listing your name first.

Name: Percent Contribution:

_____   _____

_____   _____

_____   _____

_____   _____

Describe your contribution to the *poster*.

Appendix A

Automated Data Collection

Your lab may be equipped with automated data collection devices. These are typically calculator based (CBL) or microcomputer based (MBL). Both systems have their advantages. The CBL system is more portable than the MBL system. The MBL system is directly connected to the computer so the data can be manipulated without additional data transfer steps.

The Calculator-Based Laboratory (CBL), from Vernier which sells other interfaces like the LabPro, is a device that allows you to collect and store data automatically. It functions as an interface between a probe and a TI-83 (or other Texas Instruments brand) graphing calculator. The LabPro also interfaces with a computer or PDA. There are many types of probes that can measure temperature, color intensity (absorbance), pressure, voltage, pH, and other properties.

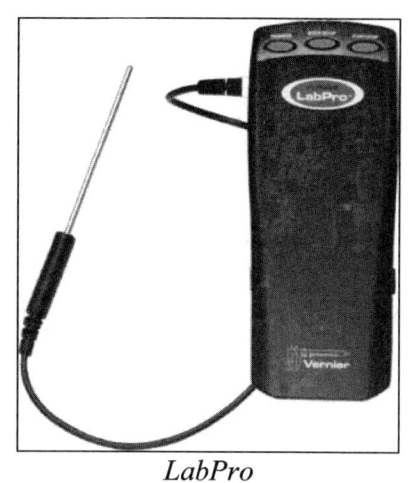

LabPro
Photograph Courtesy of Vernier Software & Technology

The calculator is equipped with programs capable of sending data collection and storage instructions to the CBL. Most of the time you will run the CHEMBIO program, which can be started by pressing [PRGM] (or [APPS] if you use a TI-83 Plus) on the calculator. If no programs appear, see your lab instructor. Use the down arrow on the calculator to highlight CHEMBIO, press [ENTER], and an introductory Vernier screen will appear. If you wish to use another program, use the down arrow key on the calculator to scroll through the other programs.

Many automated data collection systems are microcomputer-based. The interface is connected directly to a computer. Data from the probe is stored directly in the computer via the interface. The LabPro system (from Vernier) is an example of an interface that can be connected directly to a computer.

Specific instructions for your device will be provided by your instructor. Generally, when using automated data collection you must:

   a) connect the probe(s), interface, and computer (or calculator) correctly.
   b) specify the probe identity to the computer.
   c) choose the data collection mode.

The system will likely be capable of several different collection modes. Your experiment may require you to collect data manually (e.g. record the pH after the addition of base). You may want to collect data at regular time intervals (e.g. record the pH every 20 seconds for 10 minutes).

Appendix A

Learn how to use the different collection modes available to you. Automated data collection and record keeping is a very valuable tool in the chemical laboratory.

Calibration is necessary for some probes. Colorimeters and pH probes should always be calibrated immediately before use. Some probes are pre-calibrated. These pre-calibrated probes will be recognized by the computer and a stored calibration will automatically be incorporated.

After the data is collected, it may need to be transferred to a computer program of choice, and it will need to be treated for proper display for your lab report. If the data is collected with a calculator-based system, then it will need to be uploaded to a computer for adequate treatment and to ensure the printout will be professional in quality. Learn the necessary steps for getting your data into a computer with the equipment available. Microcomputer based systems will already have the data stored in the computer. Often the data is stored in a program that will manipulate the data adequately for your laboratory investigations. Many students prefer to transfer the data from the default program into a graphing software they are more comfortable with. This is usually as simple as copying the data from the default spreadsheet, opening the program of choice, and then pasting the data into the spreadsheet within. See Appendix G for more information about graphing and data processing in the laboratory.

Electronic copies of your data and graphs should always be saved to ensure they are available in case you need to modify your report.

Calculator-Based Laboratory (CBL)
Photograph Courtesy of Vernier Software & Technology

Appendix B

Transmittance and Absorbance Data Collection

Introduction

Spectroscopy is the chemical discipline that studies the interaction of electromagnetic radiation with matter. In a typical spectroscopy experiment, the item to be analyzed is placed in a chamber, then exposed to some form of electromagnetic radiation. The sample will absorb radiation at some wavelengths and other wavelengths will pass through. Spectroscopy offers many techniques that help scientists characterize substances qualitatively (identification of the species present in a material) and quantitatively (the amount of each species present). While some chemical techniques for identifying substances are destructive to the system under investigation, spectroscopic techniques are typically nondestructive to the sample of interest. The power of spectroscopy rests on the specific behavior of substances when exposed to light. For example, a certain solution will have color because light is absorbed at some wavelengths, but not at others. When light is passed through a solution, the light arrives at the wall of the container (for example, a cuvette) with some intensity, I_o. Some of the light is absorbed as it passes through the solution, so the intensity of light leaving the solution, I is less than the initial intensity.

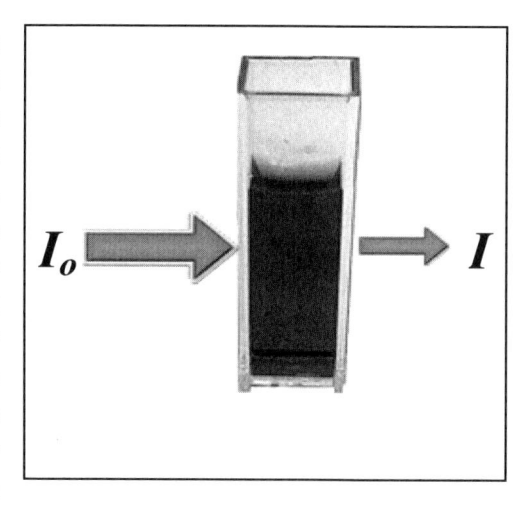

Transmittance is used to quantify the relationship between the intensity of the radiation passing through the sample (I), and the intensity of the radiation to which the sample was exposed (I_0): specifically, transmittance is the ratio of the two:

$$T = \frac{I}{I_o}$$

Or, expressed as a percentage as it often is:

$$\%T = \frac{I}{I_o} \times 100$$

When you conduct spectroscopy experiments with a spectrophotometer or colorimeter, one of the critical design issues is selection of the proper wavelength to which your sample will be exposed; that is, the light transmitted through a solution depends on the wavelengths that are absorbed by the solution components.

A detector is used to measure the intensity of the radiation passing through the sample. The detector is calibrated prior to measuring the %T of the sample. Typically the calibration is

Appendix B

achieved by assigning deionized water a value of 100%T. Additionally, the detector is blocked so that no radiation can enter, and a value of 0%T is assigned.

In addition to collecting transmittance data as described above, you can collect absorbance data. Absorbance is a measure of the amount of light absorbed by a sample. Typically a spectrometer will be capable of displaying both %T and absorbance values. The specific mathematical relationship between %T and absorbance is not shown here since it is presented as an exercise in *What Factors Affect the Intensity of Color?*

The following section describes the collection of absorbance (A) and percent transmittance (%T) data.

Transmittance and Absorbance Data Collection

When collecting absorbance or transmittance data, three steps will be involved:

a) Determining the proper wavelength of light for the sample.
b) Calibrating the spectrophotometer (or colorimeter) with an appropriate "blank".
c) Measuring the absorbance (or transmittance) of the sample(s).

Selecting the proper wavelength could be an interesting investigation by itself. The task is to find the wavelength of light where the sample has a maximum absorbance, or response, to the probe. Some colorimeters have only a few wavelengths of light available. With this type of colorimeter, a sample's absorbance could be compared to a "blank" at all available wavelengths, and the wavelength showing the maximum absorbance will be the desired wavelength. Some colorimeters have a diffraction grating which allows the user to select any wavelength within the range of the instrument. With these colorimeters, also called spectrophotometers, the user can scan through the range of wavelengths and watch the absorbance change with the varying wavelength. Again, the wavelength showing the maximum absorbance will be the desired wavelength.

Calibrating the colorimeter for a particular set of measurements will involve a few important steps. First, if necessary, the instrument must be connected to the proper interface and computer. It needs to be turned on and allowed to warm to its normal operating temperature. Calibration requires a "blank" solution. The "blank" is often deionized water. Your instrument will be calibrated by placing the "blank" in the sample chamber and telling the instrument (or computer) to read this as zero absorbance, or 100% transmittance. The instrument will also have a setting which prevents light from entering the detector. During calibration, this setting will be used and the instrument (or computer) must be told to read this as infinite absorbance, or 0% transmittance. Once the 0% T and 100% T readings have been entered into the instrument (or computer), the colorimeter is calibrated. It is important to note that the instrument can "drift" out of calibration.

The calibration settings should be checked and adjusted periodically during any investigation involving the collection of absorbance data. The instrument will need to be re-calibrated if the wavelength of light is changed.

Once the instrument is calibrated, it is ready for absorbance measurements of samples. This will often be as simple as placing the sample in the chamber, closing the cover, and reading the absorbance. If using an automated data collection device, a button or "trigger" may need to be pushed to collect and store the absorbance reading. Most CBL and MBL interfaces will collect both absorbance and percent transmittance. It is important to know which of these is being displayed by the instrument during the collection of data. Most interfaces will allow you to collect data manually or during regular assigned time intervals. Your instructor can help with the specific buttons and capabilities of the colorimeter available to you.

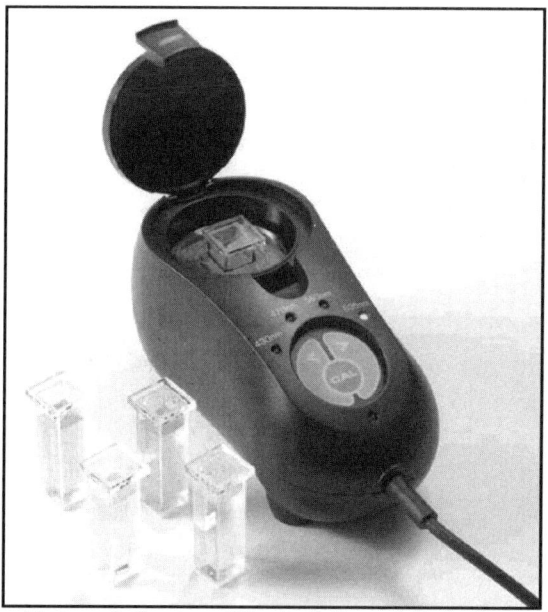

Photograph Courtesy of Vernier Software & Technology

Appendix C

Measuring pH

The solution acidity or basicity of substances is measured on a logarithmic scale called pH. Neutral water has a pH of 7, the pH of an acidic substance is less than 7, and the pH of a base is greater than 7. There are several ways to measure pH. You can set up reference solutions of known pH with universal indicator for comparison. In the presence of the indicator the solution color will vary with the pH of the solution. Alternatively, you could also use a pH meter or pH paper. The method you use depends on how accurately you want to know the pH.

Indicators

Colored indicators are often used to measure pH. These indicators are a different color in a basic environment than in an acidic environment. Litmus, for example, is blue in a basic environment and red in an acidic environment. Paper coated with litmus (litmus paper) can be used to determine whether a solution is basic or acidic. Paper coated with Universal Indicator (pH paper) can be dipped into a solution to determine its pH within 1 pH unit. Universal Indicator is a combination of indicators which will produce a different color for each pH value. To use pH paper, just dip the paper into the solution to be measured and then compare the color to the color chart supplied with the paper. To measure the pH of a gas, moisten the pH paper with deionized water and then expose the moistened paper to the gas.

An indicator can be added directly to a solution to determine its pH. The color change for an indicator occurs at a specific pH range. The specific pH range may be supplied with your indicator, or may only be available to you by using a reference book. If you know the pH range at which an indicator changes color, then you can know whether a solution is at that pH, above that pH, or below that pH.

pH Meter

A pH meter consists of two electrodes connected to a voltmeter, which is capable of measuring small differences in electrical potential. A pH meter can often measure pH to within 0.01 pH units. Most manufacturers make combination electrodes which have both electrodes physically located within one tube. One electrode is a reference and the other (the indicator electrode) is sensitive to changes in pH. The most common indicator electrode is the glass electrode which contains a solution within a glass bulb having a thin glass membrane. If the solution on the outside of the membrane is of a different pH than the solution within, then an electrical potential will develop across the membrane. This potential difference is measured by the voltmeter. A pH meter will directly convert the voltage into pH units. The meter must be calibrated prior to making pH measurements. A CBL or MBL system equipped with a pH probe can be used as a pH meter and must similarly be calibrated.

Appendix C

This appendix describes the procedure for calibrating and making pH measurements. Directions for use will vary slightly with the model of pH meter used. Consult your instructor or the operator's manual for specific instructions for your pH meter or probe.

General Instructions for pH Measurement

1. Electrodes should always be supported with a clamp when in use. If separate electrodes are used, then they should not be allowed to touch each other or the bottom or sides of the container.

2. Glass electrodes are extremely fragile. Be careful. If the electrode is covered with a guard sleeve, then do not remove this sleeve. Make sure the electrode is immersed far enough to cover the perforation in the sleeve when making measurements.

3. After making a pH measurement, immediately rinse the electrode(s) with deionized water and gently pat dry with a lint-free tissue to prevent contamination of the next solution.

4. When the pH meter is not in use, the electrodes must be soaked in water or a neutral pH buffer solution. If the glass membrane dries out, then the electrode will become inoperative.

5. When calibrating the pH meter, use at least two standard buffer solutions. Ideally, one should have a pH value higher than the solutions to be measured, and the other should have a pH value lower than the solutions to be measured. It is best if the buffers are also close in pH to the solutions of interest.

Taking pH Measurements in the Laboratory

1. Connect the pH probe to the appropriate meter, or interface and computer.

2. Select the appropriate software (if necessary) and identify the probe type (pH).

3. You should always perform new calibrations with the pH probe. When using a pH probe and interface, the software will take you through a two-point calibration procedure for which buffer solutions are needed. When the software prompts you to calibrate point 1, insert the pH probe in the lower pH buffer solution, wait until the voltage reading stabilizes, and import the data (this may require you to push a button or "trigger").*

4. The software will then prompt you for the pH of the buffer solution; input the value to assign the proper pH to the imported data.

5. You will then be prompted for the second reference point. After rinsing the probe in deionized water and patting the tip dry with a soft tissue, insert the probe in the higher pH buffer solution. Again, wait until the voltage reading stabilizes, and import the data (this may require you to push a button or "trigger").

6. Input the pH of the second pH reference solution. You can now collect pH data.

Measuring pH

7. Select the desired mode of data collection. Most interfaces will allow you to collect data manually or during regular assigned time intervals. Your instructor can help with the specific buttons and capabilities of the pH system available to you.

8. After rinsing the probe in deionized water and again patting the tip dry with a soft tissue, insert the probe into the solution to be measured and start your manual or automated data collection.

9. When you are finished using the pH probe, rinse it and immerse it in its storage solution.

*A laboratory pH meter is a self-contained instrument that includes a pH probe, interface, voltmeter, and a display or readout. This instrument is typically calibrated by "dialing" the instrument's standardization knob to match the pH's of the buffer solutions. Your instructor can help with the specific buttons and steps required to calibrate your pH meter if one is available to you.

Appendix D

Temperature Data Collection

Temperature can be measured using a thermometer or a temperature probe connected to a CBL or MBL system. The automated systems are particularly useful when a series of temperature measurements are required over the course of an experiment. Some brief instructions for using a temperature probe with the CBL are listed below. The MBL system would probably require less steps than the CBL system, and would have the added advantage of storing your temperature data directly in the computer. For more information about CBL and MBL systems, see Appendix A.

Temperature Measurements with the CBL or MBL

1. Connect the temperature probe to the interface unit and to the calculator (or computer).

2. Select the appropriate software (if necessary) and identify the probe type (temperature). The software may already recognize the probe type without any additional identification. Because the temperature probe is already calibrated, a calibration menu will probably not appear when you use the temperature probe.

3. Select the desired mode of data collection. Most interfaces will allow you to collect data manually or during regular assigned time intervals. Your instructor can help with the specific buttons and capabilities of the automated data collection system available to you.

Appendix E

Pressure Data Collection

Pressure changes can be monitored using a pressure sensor connected to an interface and a graphing calculator, or connected to an MBL system. Pressure can also be measured in the laboratory using a simple manometer. This appendix shows some details of taking pressure measurements with a CBL or MBL system, and describes the use of a simple laboratory manometer. Consult your instructor if your equipment differs significantly from that described here.

Pressure Measurements with a CBL and Pressure Sensor

Figure 1
(Photograph Courtesy of Vernier Software & Technology)

1. Prepare the pressure sensor for data collection.

   - Plug the pressure sensor into the interface and calculator (or computer). A 20-mL syringe or other gas container can be connected to the valve.
   - To allow air or other gases to enter or exit the pressure sensor and container, open the valve to the pressure sensor. Some pressure sensors have a three-way valve. This type valve is used to select which gas sample is to be exposed to the probe. The probe can be open to the room, open to a sample chamber and closed to the room, or closed to both (see Figures 2 and 3). Familiarize yourself with the valve connected to your pressure sensor.

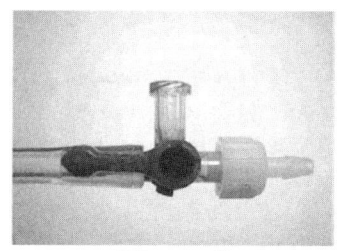

Figure 2

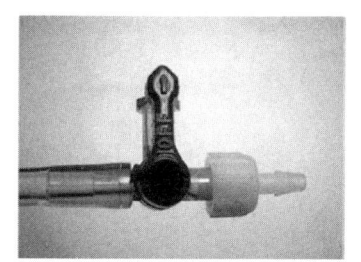

Figure 3

2. Select the appropriate software (if necessary) and identify the probe type (pressure).

3. Select the desired mode of data collection. Most interfaces will allow you to collect data manually or during regular assigned time intervals. Your instructor can help with the specific buttons and capabilities of the pressure system available to you.

4. Collect your data. Some pressure sensors only measure pressures up to 1.5 atm. Some laboratory equipment will break or explode when gases are produced or heated in a closed system. Know the limits of your equipment and the pressure probe before designing any investigations involving pressure measurements.

Appendix E

Using a Simple Manometer

A manometer can be used to measure pressure changes in the laboratory. A manometer consists of a U-shaped tube filled with a liquid (usually water or mercury). One end of the tube is connected to the chamber where the pressure is to be monitored, and the other end is open to the lab.

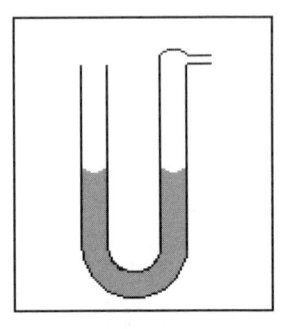

The height difference of the liquid on either side of the tube (Δh) is directly proportional to the pressure difference between the chamber and the lab (ΔP). The pressure difference can be measured with a ruler and be recorded in units of mmHg or mmH$_2$O. Often the actual pressure of the lab (atmospheric pressure) can be measured with a barometer. Consult your instructor to find out whether a barometer is available to you. If you know the atmospheric pressure and ΔP, then the pressure of the gas in the chamber can be calculated using simple addition or subtraction.

If your manometer contains mercury, be careful. If the pressure in the chamber increases dramatically, there is a danger that the mercury could be ejected from the tube into the lab. (Note that most mercury manometers have a "safety trap" to prevent that from happening.)

Appendix F

Voltage Data Collection

The potential difference between two processes (for example, half–reactions) can be measured using a voltmeter. One wire from the voltmeter is attached to one half–reaction, another wire from the voltmeter is attached to the other half–reaction, and the voltage is read on a digital or other type readout. Voltmeters typically do not require calibration prior to taking measurements. Voltmeters are inexpensive and readily available. A CBL or MBL equipped with a voltage probe can function as a voltmeter.

Familiarize yourself with the voltmeter before taking measurements. Make sure you are measuring voltage and not some other property. Many voltmeters are capable of measuring resistance and current (amps) as well as voltage. Often you must select "voltage" or "volts" to specifically measure voltage. Consult your instructor or the operator's manual if you are not clear about how to measure voltage with your voltmeter.

Voltage Measurements with a Voltage Probe

To use the voltage probe, follow the same procedure for using the temperature probe except identify the "voltage" probe. Like the temperature probe, the voltage probe may already calibrated so a calibration menu will probably not appear. After selecting the voltage probe, you can collect voltage data using the data collection mode appropriate for your experiment.

Photographs Courtesy of Vernier Software & Technology

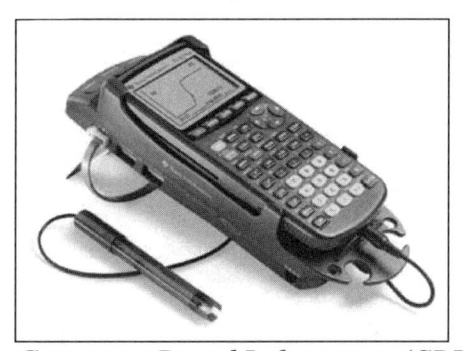

Computer-Based Laboratory (CBL)   *LabPro*

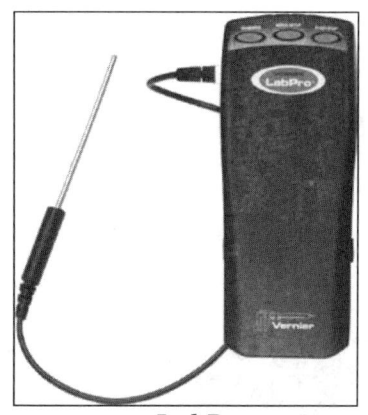

If a negative voltage reading is observed, then just reverse the leads. You should get the same numerical value as a positive voltage. You should also note that when the voltage probe is not connected to anything, a default voltage reading of 2.0 V may be displayed with the CBL system.

Voltage Data Collection

Appendix G

Selected Laboratory Techniques

This appendix describes techniques commonly used in the chemistry laboratory. Included are the proper ways to use a balance, a pipet, a buret, a graduated cylinder, a Bunsen burner, filtration equipment, and a centrifuge. There is also information about titrations, graphing, and data processing.

Taking Mass Measurements

One very important rule to follow when taking mass measurements is **never put chemicals directly on the balance pan.** This can damage the balance. Always use weighing paper when measuring the mass of a solid, and of course, pour liquids into a preweighed container before measuring their mass.

Mass measurements are usually taken using either a triple beam balance or an electronic balance. The electronic balance has several advantages over the triple beam balance, but is significantly more expensive. If your balance is not like those described here, consult your instructor for directions on the proper use of your balance.

Triple Beam or Platform Balance

Zero the balance by adjusting the knob at one end of the balance. The adjustment knob is usually just a small weight that can be moved along a screwlike shaft. Check the zero point periodically since foreign matter may accumulate on the platform or on the beams and cause a slight change in the zero position.

After the balance has been zeroed, place the sample on the platform (remember to use weighing paper or an appropriate container) and move the large weight to the notch **just before** the notch which causes the pointer to drop. Then move the next weight to the right in the same fashion. Finally slide the small weight along the scale until the pointer rests at zero. The mass is the sum of all of the weight positions, read directly from the three beams.

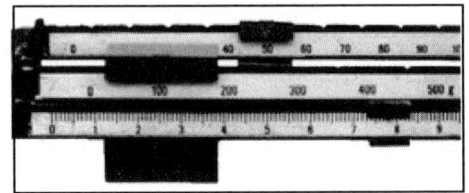

Electronic Balance

You should check to see that the balance is level before use. The balance should have a "leveling bubble." Adjust the length of each leg of the balance until the "leveling bubble" is in the center of its circular chamber.

An electronic balance is typically very easy to use. Simply turn the balance on and "tare." Taring the balance means zeroing the balance. After pushing the tare button, the balance should read 0.000 g. The last digit may fluctuate because of air currents in the room. This is expected. After taring the balance, place the sample on the pan (remember to use weighing paper or an appropriate container) and then read the mass directly from the balance display. It's very easy. Turn the balance off when finished.

The tare button can be used to measure the mass of a solid or liquid directly. Just place the weighing paper (or container) on the balance and tare. The paper (or container) now will read as 0.000 g. Then remove the paper (or container), place the desired solid (or liquid) into it, and return it to the balance pan. The mass shown on the balance display is now the mass of the solid (or liquid). **Never pour a liquid into a container while it is on the balance.**

Measuring Volume

Volume measurements are typically made with a graduated cylinder, pipet, buret, or volumetric flask. A beaker or Erlenmeyer flask can be used if accuracy is not important. Some volumetric devices are designed to accurately measure the volume of liquid **contained** and some devices are designed to accurately measure the volume of liquid **delivered** by it. A device marked with "TD" is designed to measure the amount delivered and a device marked with "TC" is designed to measure the volume of liquid contained. This section will describe the proper use of a graduated cylinder, pipet, buret, and volumetric flask.

Note that it is important to report your volume measurements with the correct number of significant figures. The number of significant figures varies with the type of equipment.

Appendix G

Graduated Cylinder

When reading the volume using a graduated cylinder, always read the bottom of the meniscus. Estimate one digit between the nearest two graduations. For example, if your graduated cylinder is marked with 1 mL graduations, and the bottom of your meniscus is between the 16 mL and 17 mL marks, and you estimate it's $^3/_{10}$ of the way between 16 and 17, then you must report the volume as 16.3 mL.

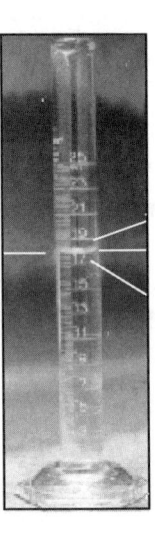

Correct viewpoint

Pipet

A volumetric pipet is designed to **deliver** an accurate volume of liquid. Some pipets have graduations and some have only one line for measuring volume. It takes practice to use a pipet properly. Allow time for practice whenever carrying out an investigation that involves the use of a pipet.

If your pipet has graduations, then you can accurately measure the volume of liquid delivered by the pipet by taking a measurement before and one after your delivery, and reporting the difference as the volume delivered. The number of significant figures will depend on the markings. You should always estimate one digit as with the graduated cylinder (discussed earlier).

If your pipet has only one line for measuring volume, then this line represents a volume delivery accurate to four significant figures. For example, if a 25 mL pipet is filled so that the bottom of the meniscus is just touching that line, then 25.00 mL of volume will be delivered upon draining that pipet. Note that a pipet is not really emptied when used properly.

To use a pipet,

1. Use a bulb to draw the liquid to a level above the desired amount (above the mark). **Never draw a liquid by sucking the end of the pipet.**

2. Remove the bulb and quickly place your index finger over the end of the pipet. By adjusting your finger pressure, allow the liquid to drain to the line. Sometimes you can slow the flow of the liquid by touching the pipet tip to the bottom of the container.

3. Remove the pipet from the solution (keeping your finger pressure on the end) and remove any lingering drops from the tip by touching the tip to the side of the container. Wipe off any liquid on the outside of the pipet with a paper towel.

4. To deliver the liquid, remove your finger from the pipet end. After draining, touch the tip to the side of the receiving vessel. You have now delivered a volume accurate to four significant figures.

You may notice there is still some liquid in the pipet tip. That's okay. If you use a bulb to "blow" that extra liquid into the receiving vessel, then you will deliver too much liquid.

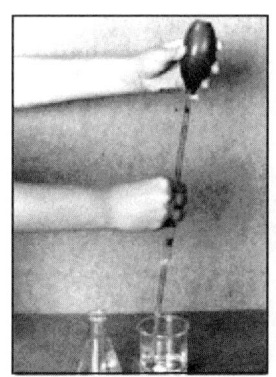

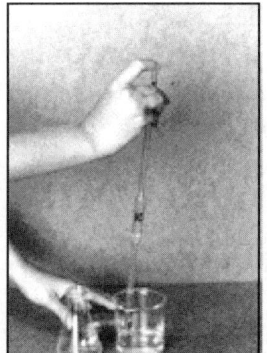

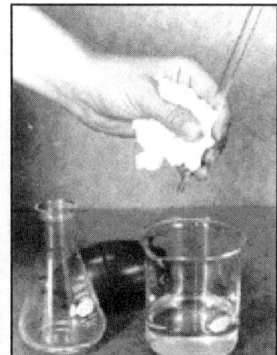

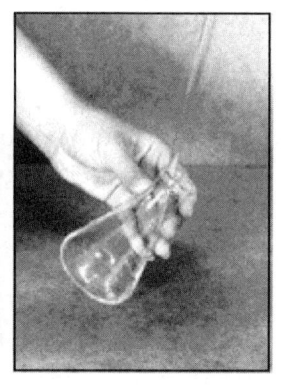

Buret

A buret is designed to deliver a volume accurate to ± 0.01 mL. The buret should be secured in a proper buret stand when in use. A delivered volume is measured by simply taking a reading from the bottom of the meniscus before and after the delivery. The difference in readings is the volume delivered. The delivery is accomplished by opening the stopcock at the bottom of the buret. The flow rate can be controlled by adjusting the stopcock position. Most laboratory burets hold 50 mL and are marked in tenths of a mL.

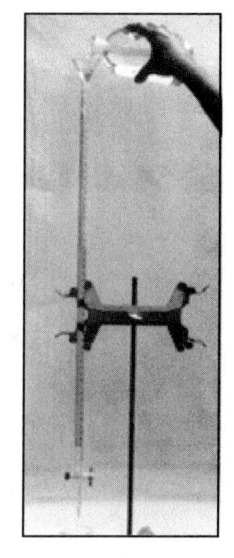

Selected Laboratory Techniques

Appendix G

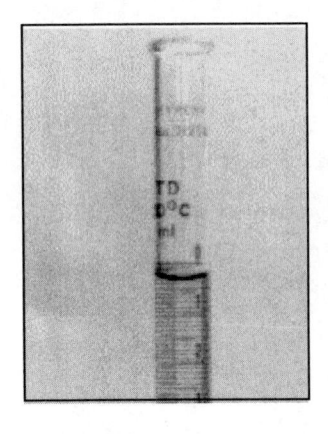

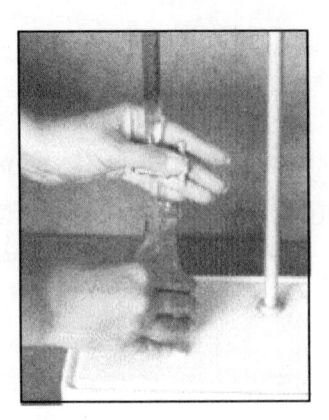

You should always rinse your buret with a few portions of the solution prior to filling. It is good practice to use a funnel when filling the buret, and to hold the buret below eye level when filling. It is important to remove all air bubbles from the buret tip before using the buret to make measurements. This is usually accomplished by allowing the liquid to flow through the tip until all of the air bubbles are blown out.

Volumetric Flask

A volumetric flask is designed to contain an accurate volume of liquid. It is normally used to prepare a solution where a final volume measurement is needed. A volumetric flask has one volume mark on the neck of the flask. A typical volumetric flask is assumed to be accurate to four significant figures. For example, if a 50 mL volumetric flask is filled to the mark, then it contains 50.00 mL.

To use a volumetric flask,

1. Select the volume desired. Volumetric flasks come in many volumes. The most common are 10, 25, 50, 100, 250, 500, and 1000 mL.
2. Place a measured amount of solute, or solution to be diluted, into the clean volumetric flask.
3. Add deionized water to the flask until it is about ¾ filled, and dissolve the solute by swirling.
4. Add more water until the meniscus gets close to the mark on the neck of the flask. Then add water drop by drop until the bottom of the meniscus just touches the mark.

Selected Laboratory Techniques

5. Stopper and invert the flask several times to ensure complete mixing. You have now prepared a solution with a volume measurement accurate to four significant figures.

Using a Bunsen Burner

A Bunsen burner produces a flame for heating things in the laboratory. The burner must be connected to a source of natural gas (methane). The burner will typically have a valve to control the gas flow rate. Most also have an airflow control. The gas and air are mixed in the Bunsen burner and ignited to produce the desired flame.

To light the burner, adjust the air control so that a minimum of air is supplied, turn on the gas, and hold a lighted match (or other lighter) above and to one side of the gas flow. After ignition, increase the airflow until a blue flame is obtained. The hottest part of the flame is just above the inner blue cone of the flame. If the blue cone is absent, increase the air flow. If the flame rises above the burner, turn down the gas flow. If the flame blows out, turn down the air flow. Never leave a flaming Bunsen burner unattended.

When heating a glass vessel such as a test tube, make sure you are using Pyrex or Kimax glassware, which can withstand the intense heat. When heating a test tube, always use a test tube holder and don't point the opening of the test tube at anyone. When heating other vessels, use a ring stand, ring, and wire gauze to hold the vessel above the burner.

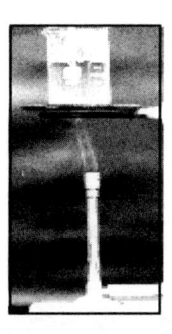

Filtration (Separating a Solid and a Liquid)

A solid and a liquid can be separated using filtration or centrifugation. This section describes the techniques of filtration.

Gravity Filtration

1. Fold a circular piece of filter paper in half. Fold in half again.
2. Open the paper to form a cone and fit the cone into a funnel.
3. Wet the paper with a few milliliters of deionized water (or other appropriate solvent) and press the cone to the sides of the funnel so it fits tightly.
4. Pour the mixture into the cone. A clear solution should come through the filter and run through the funnel into a clean container.

Appendix G

Avoid filling the filter too full. It is desirable to have the stem of the funnel on the side of the receiving container so the liquid filtrate doesn't splatter. Make sure the funnel stem is not submerged in liquid. This could prevent the flow of liquid through the filter.

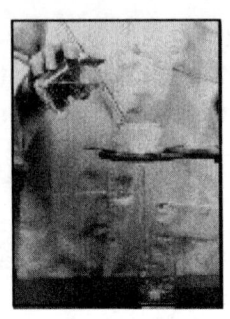

Suction Filtration

1. Place a piece of filter paper flat against a Büchner funnel bottom and moisten with deionized water (or other appropriate solvent) to seal.
2. Place the funnel onto a side-arm flask with an appropriate vacuum seal.
3. Attach a vacuum line. A vacuum can be created by flowing water from a faucet through an aspirator.
4. Turn on the vacuum (or aspirator).
5. Pour the mixture into the Büchner funnel. The solution should be pulled through the funnel.
6. The solid residue should be washed with deionized water. If deionized water could re-dissolve the solid, then wash it with another solvent. If the liquid filtrate is to be kept, then don't wash the solid residue until after the filtrate has been removed from the side-arm flask.
7. The solid can be dried by allowing air to be pulled through it by the vacuum.
8. Make sure to disconnect the apparatus before you turn off the vacuum or aspirator. Otherwise, if you are using an aspirator, you could potentially draw tap water from the aspirator into your flask.

Using a Centrifuge

A centrifuge is a device used to separate a solid from a liquid by centrifugal force. It usually consists of four to eight openings in a rotor head. The rotor head spins at high speed, causing the usually more dense solid to be driven to the bottom of the test tube containing the mixture. The liquid can then be removed by decanting (pouring) or by withdrawing with a capillary pipet.

To use a centrifuge:

1. Place the sample to be separated into a test tube that will fit into the rotor head openings.
2. Pour deionized water into another test tube (same size) to the same depth as the mixture in the sample tube. This tube will act as a counterweight to keep the centrifuge balanced.
3. Place the test tubes into the rotor openings opposite one another.
4. Turn on the centrifuge and allow it to attain full speed.
5. Allow the centrifuge to run. Some mixtures separate in about one minute. Others can take up to eight minutes.
6. Switch off the centrifuge and allow it to stop without assistance. It's dangerous to touch the rotor while it's spinning.
7. Remove both test tubes.
8. If the solid is separated from the liquid, pour the liquid out (this is called decanting). The liquid solution should be free of solid particles.
9. The solid is still contaminated with the solution and should be washed. Add a few mL of deionized water to the solid and mix well. Centrifuge and decant.
10. Repeat step 9 two more times. Your solid should now be free of contamination from the dissolved ions.

Titration

A titration is used to determine the concentration of a solution by reacting it with a measured amount of another solution of known concentration. The solution of known concentration is called the **standard solution**. A carefully measured volume of one solution is placed in an Erlenmeyer flask along with a few drops of an indicator solution. The other solution is slowly added from a buret into the swirling Erlenmeyer flask until the stoichiometric equivalence point is reached. This is observed by a change in color of the indicator. It is also possible to detect an equivalence point using a pH meter or other device.

Once the two solution volumes are known, the concentration of the solution can be calculated using these volumes and the concentration of the standard solution. For example, if 25.00 mL of $H_2SO_4(aq)$ (unknown concentration) were pipeted into an Erlenmeyer flask, and 22.08 mL of 0.1035 M NaOH(aq) were added to reach the equivalence point, then the $H_2SO_4(aq)$ concentration can be calculated as follows:

First calculate the number of moles of NaOH delivered during the titration:

$$[0.02208 \text{ L NaOH (aq)}] \times \frac{0.1035 \text{ mol NaOH}}{1 \text{ L NaOH (aq)}} = 2.285 \times 10^{-3} \text{ mol NaOH}$$

Then, using the balanced equation, calculate the number of moles of H_2SO_4 that must have reacted with the added NaOH:

$$2NaOH(aq) + H_2SO_4(aq) \rightarrow 2H_2O(l) + Na_2SO_4(aq)$$

$$[2.285 \times 10^{-3} \text{ mol NaOH}] \times \frac{1 \text{ mol } H_2SO_4}{2 \text{ mol NaOH}} = 1.143 \times 10^{-3} \text{ mol } H_2SO_4$$

Finally, determine the mole/L ratio for the H_2SO_4 solution knowing the moles of H_2SO_4 that reacted and the volume of $H_2SO_4(aq)$ used:

$$\frac{1.143 \times 10^{-3} \text{ mol } H_2SO_4}{0.02500 \text{ L } H_2SO_4 \text{ (aq)}} = 0.04571 \, M \, H_2SO_4$$

The final step of your calculation may be to determine the grams of acid consumed, or some other quantity; but the theme is always the same. A reaction is carried out to its stoichiometric equivalence point, and the moles of solute that reacted from a solution is determined by knowing the moles of solute that reacted from the standard solution, and knowing the mole:mole ratio in which the two substances react.

Graphing and Data Processing

Data tables and graphs are used in scientific reports because they can convey a great deal of information about an experiment in a concise fashion. This section will give some general guidelines about reporting data and graphing, and will point out the numerous advantages of computer-aided graphing.

Guidelines:

1. Graphs and tables should always have an appropriate title.
2. Graphs and tables should always clearly report the units for each numerical entry. On a graph, the axes should be labeled with the appropriate variable name, and the units. Table entries should contain units, or should have the units clearly shown in the table heading.
3. If a table entry is a calculated value, then the calculation should be shown as a footnote to the table. For example, if the mass of a container is in row 1, and the mass of the container with solid is in row 2, and the mass of the solid is in row 3, and the mass of the solid was found from the other two values, then an appropriate footnote to the table would be:

$$\text{mass solid} = (\text{mass container with solid}) - (\text{mass container})$$

4. Graphs should be large enough to show any trends in the data. It is usually preferable for the graph to be scaled so that the data fill the entire graphing area, and for the graph to fill an entire page.

Graphing

Graphs can be generated by hand using graph paper, or by using a computer equipped with graphing software such as Graphical Analysis, Excel, CricketGraph, or any other similar software. When using a computer, many of the steps required for producing a good graph are done automatically.

Graphing by Hand

When graphing by hand, it is important to scale the x and y axes such that all of the data are included and they fill the graphing area. For example, if the x values range from 0 to 56, then your x axis would range from 0 to 60. If your y values range from 2×10^{-2} to 8.6×10^{-1}, then your y axis might range from 1×10^{-2} to 1.

Once your x and y axis ranges are determined, label the axes with their appropriate variable names and units. Enter x and y values at even intervals along the axes to show the range along each axis. Plot the points using a sharp pencil.

If the data show, or are supposed to show a linear trend, it would be appropriate to draw a straight line through the data points, and determine the slope and y intercept of the line. Often the data are not in a perfectly straight line, and you will need to draw the best straight line you can while considering all of the data points. The line may touch several points, but probably not all of them. Ask your instructor to help if you need to draw a straight line through data points that are not in a straight line. Once the line is drawn, you can calculate the slope by picking two convenient points **along the line** (not necessarily data points) and dividing the difference in y values by the difference in x values:

Appendix G

$$\text{slope } (m) = \frac{(y_1 - y_2)}{(x_1 - x_2)}$$

This calculation may give a negative number, corresponding to a declining slope; or a positive number, corresponding to a rising slope. The *y* intercept is determined by observing where the line crosses the y axis. This is the y value when $x = 0$. You can also determine the *y* intercept value mathematically using the equation of the line:

$$y = mx + b$$

If you have a y value, its corresponding *x* value, and the slope of the line (*m*), you can solve the equation for the y intercept (*b*).

Graphing with a Computer

To graph with a computer, the data must be imported (or typed) into a graphing software program such as Graphical Analysis. The Graphical Analysis software allows direct importing of data from a graphing calculator (under the File menu). Once your data is stored on the computer, it can probably (depending on the age and model of your computer) be simply highlighted and cut and then pasted into the graphing program of your choice.

The graphing program will show your imported (or typed) data in columns. At some point you will need to assign labels for each column of data.

Graphing is often as simple as pushing a button and choosing which variable you want on the x axis and which variable you want plotted on the y-axis. You can also choose to plot multiple data sets on the same graph. After making your choices, the graph will appear automatically scaled. At this point you can usually assign a title, edit the axis labels, adjust the scale, put in gridlines, ask the computer to draw the best straight line fit and report the slope and intercept, view the graph as it would look on a page when printed, or just about anything else you could imagine to help you create a professional–quality graph.

The exact steps you take to polish your graph will vary from program to program. Familiarize yourself with the graphing program available to you and take advantage of its tremendous capability.

If you need to do a mathematical operation on your data, then graphing and spreadsheet programs easily allow you to write a formula in another column that carries out the calculation. Suppose you have a column of Celsius temperature data that you want to convert to Fahrenheit. Your software is probably capable of accepting the mathematical formula for converting the data. The software will then automatically convert the entire column into the converted values, and display the converted data in a new column. This is a very valuable tool for laboratory investigations. You should know how to convert and/or manipulate columns of data using the graphing software available to you.

Appendix H

Laboratory Equipment

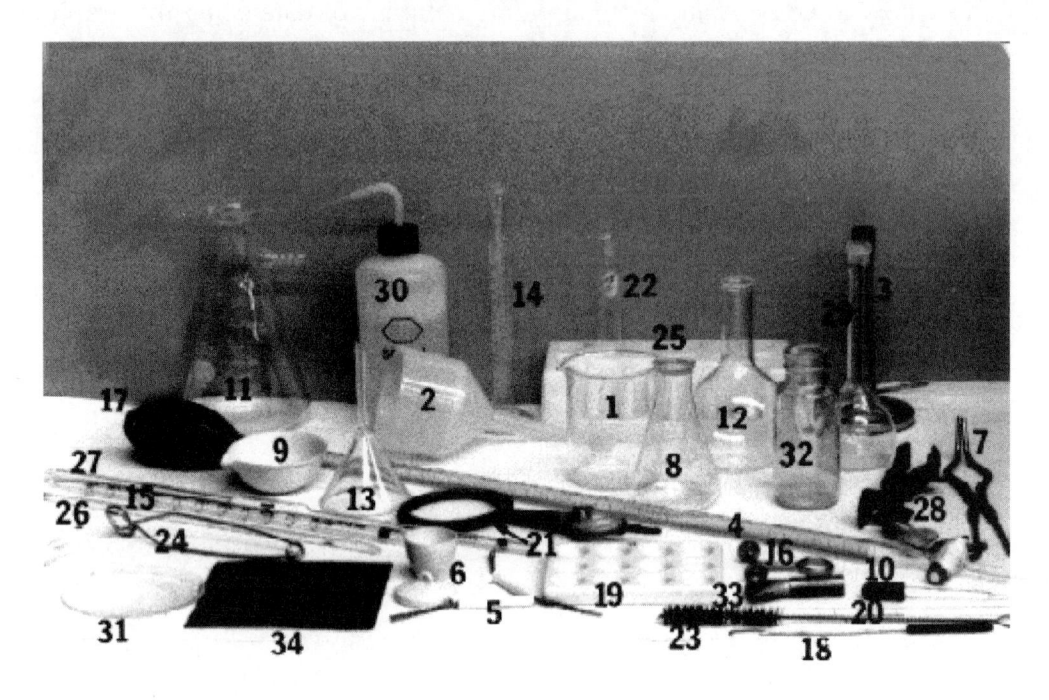

1 – beaker
2 – Büchner funnel
3 – Bunsen burner
4 – buret
5 – clay triangle
6 – crucible and lid
7 – crucible tongs
8 – Erlenmeyer flask
9 – evaporating dish
10 – eye dropper
11 – filter flask
12 – Florence flask
13 – funnel
14 – graduated cylinder
15 – measuring pipet
16 – pinch clamp
17 – pipet bulb
18 – spatula
19 – spot plate
20 – stirring rod
21 – support ring
22 – test tube
23 – test tube brush
24 – test tube holder
25 – test tube rack
26 – thermometer
27 – transfer pipet
28 – utility clamp
29 – volumetric flask
30 – wash bottle
31 – watch glass
32 – wide mouth bottle
33 – wing top
34 – wire gauze

Sample Material Safety Data Sheet

Acetone
00140

**** SECTION 1 - CHEMICAL PRODUCT AND COMPANY IDENTIFICATION ****

MSDS Name: Acetone
Synonyms:
  Dimethylformaldehyde, dimethyl ketone, 2-propanone, pyroacetic acid, pyroacetic ether
Company Identification:

For information, call:
Emergency Number:
For CHEMTREC assistance, call:

**** SECTION 2 - COMPOSITION, INFORMATION ON INGREDIENTS ****

CAS#	Chemical Name	%	EINECS#
67-64-1	2-propanone	99	200-662-2

**** SECTION 3 - HAZARDS IDENTIFICATION ****

EMERGENCY OVERVIEW

Appearance: Colorless, highly volatile liquid. Flash Point: -4°F. Danger! Extremely flammable liquid. May cause central nervous system depression. May cause liver and kidney damage. Causes eye and skin irritation. Causes digestive and respiratory tract irritation. Target Organs: Kidneys, central nervous system, liver, respiratory system.

Potential Health Effects
  Eye:
    Produces irritation, characterized by a burning sensation, redness, tearing, inflammation, and possible corneal injury.
  Skin:
    Exposure may cause irritation characterized by redness, dryness, and inflammation.
  Ingestion:
    May cause irritation of the digestive tract. May cause central nervous system depression, kidney damage, and liver damage. Symptoms may include: headache, excitement, fatigue, nausea, vomiting, stupor, and coma.
  Inhalation:
    Inhalation of high concentrations may cause central nervous system effects characterized by headache, dizziness, unconsciousness and coma. Causes respiratory tract irritation. May cause liver and kidney damage. May cause motor incoordination and speech abnormalities.
  Chronic:
    Prolonged or repeated skin contact may cause dermatitis. Chronic inhalation may cause effects similar to those of acute inhalation.

Source: Fisher Scientific MSDS Web site, http://www.fisher1.com.

Appendix I

**** SECTION 4 - FIRST AID MEASURES ****

Eyes:
    Flush eyes with plenty of water for at least 15 minutes, occasionally lifting the upper and lower lids. Get medical aid immediately.

Skin:
    Flush skin with plenty of soap and water for at least 15 minutes while removing contaminated clothing and shoes. Get medical aid if irritation develops or persists.

Ingestion:
    If victim is conscious and alert, give 2-4 cupfuls of milk or water. Get medical aid immediately.

Inhalation:
    Get medical aid immediately. Remove from exposure to fresh air immediately. If not breathing, give artificial respiration. If breathing is difficult, give oxygen.

Notes to Physician:
    Treat symptomatically and supportively.
    No specific antidote exists.

**** SECTION 5 - FIRE FIGHTING MEASURES ****

General Information:
    Containers can build up pressure if exposed to heat and/or fire. As in any fire, wear a self-contained breathing apparatus in pressure-demand, MSHA/NIOSH (approved or equivalent), and full protective gear. Vapors can travel to a source of ignition and flash back. Use water spray to keep fire-exposed containers cool.

Extinguishing Media:
    For small fires, use dry chemical, carbon dioxide, water spray or alcohol-resistant foam. For large fires, use water spray, fog, or alcohol-resistant foam.

Autoignition Temperature: 33°F (0.56°C)
Flash Point: -4°F (-17.78°C)
NFPA Rating: health-1; flammability-3; reactivity-0
Explosion Limits, Lower: 2.5
                Upper: 12.8

**** SECTION 6 - ACCIDENTAL RELEASE MEASURES ****

General Information: Use proper personal protective equipment as indicated in Section 8.

Spills/Leaks:
    Absorb spill with inert material, (e.g., dry sand or earth), then place into a chemical waste container. Wear appropriate protective clothing to minimize contact with skin. Remove all sources of ignition.

**** SECTION 7 - HANDLING and STORAGE ****

Handling:
    Wash thoroughly after handling. Remove contaminated clothing and wash before reuse. Use with adequate ventilation. Avoid contact with eyes, skin, and clothing. Empty containers retain product residue, (liquid and/or vapor), and can be dangerous. Do not pressurize, cut, weld, braze, solder, drill, grind, or expose such containers to heat, sparks or open flames.

Storage:
    Keep away from sources of ignition. Store in a tightly closed container.

**** SECTION 8 - EXPOSURE CONTROLS, PERSONAL PROTECTION ****

Engineering Controls:
Use process enclosure, local exhaust ventilation, or other engineering controls to control airborne levels below recommended exposure limits.

Exposure Limits

Chemical Name	ACGIH	NIOSH	OSHA - Final PELs
2-propanone	750 ppm; 1780 mg/m3; 1000 ppm STEL; 2380 mg/m3 STEL	250 ppm TWA; 590 mg/m3 TWA	1000 ppm TWA; 2400 mg/m3 TWA

OSHA Vacated PELs:
2-propanone:
750 ppm TWA; 1800 mg/m3 TWA; 1000 ppm STEL; 2400 mg/m3 STEL

Personal Protective Equipment
   Eyes:
      Wear chemical goggles and face shield.
   Skin:
      Wear appropriate gloves to prevent skin exposure.
   Clothing:
      Wear polyethylene gloves, apron, and/or clothing.
   Respirators:
      Follow the OSHA respirator regulations found in 29CFR 1910.134. Always use a NIOSH-approved respirator when necessary.

**** SECTION 9 - PHYSICAL AND CHEMICAL PROPERTIES ****

Physical State:	Liquid
Appearance:	Colorless, highly volatile liquid.
Odor:	Sweetish.
pH:	7
Vapor Pressure:	180 mm Hg
Vapor Density:	2.0 (Air=1)
Evaporation Rate:	7.7 (n-Butyl acetate=1)
Viscosity:	Not available
Boiling Point:	133.2°F
Freezing/Melting Point:	-139.6°F
Decomposition Temperature:	Not available.
Solubility:	Not available.
Specific Gravity/Density:	0.79 (Water=1)
Molecular Formula:	C_3H_6O
Molecular Weight:	58.0414

**** SECTION 10 - STABILITY AND REACTIVITY ****

Chemical Stability:
   Stable.
Conditions to Avoid:
   High temperatures, temperatures above 220°C.
Incompatibilities with Other Materials:
   Forms explosive mixtures with hydrogen peroxide, acetic acid, nitric acid, nitric acid+sulfuric acid, chromic anhydride, chromyl chloride, nitrosyl chloride, hexachloromelamine, nitrosyl perchlorate, nitryl perchlorate, permonosulfuric acid, thiodiglycol+hydrogen peroxide.
Hazardous Decomposition Products:
   Carbon monoxide, carbon dioxide.
Hazardous Polymerization: Has not been reported.

**** SECTION 11 - TOXICOLOGICAL INFORMATION ****

RTECS#:
   CAS# 67-64-1: AL3150000
LD50/LC50:
   CAS# 67-64-1: Inhalation, rat: LC50 =50100 mg/m3/8H; Oral, mouse: LD50 = 3 gm/kg; Oral, rabbit: LD50 = 5340 mg/kg; Oral, rat: LD50 = 5800 mg/kg; Skin, rabbit: LD50 = 20 gm/kg.
Carcinogenicity:
   2-propanone -
   Not listed by ACGIH, IARC, NIOSH, NTP, or OSHA.
Epidemiology:
   No information available.
Teratogenicity:
   No information available.
Reproductive Effects:
   Fertility: post-implantation mortality. Ihl, mam: TCLo=31500 ug/m3/24H (1-13D preg)
Neurotoxicity:
   No information available.
Mutagenicity:
   Cytogenetic analysis: hamster fibroblast, 40 g/L Sex chromosome loss/non-disjunction: S.cerevisiae, 47600 ppm
Other Studies:
   None.

**** SECTION 12 - ECOLOGICAL INFORMATION ****

Ecotoxicity:
   Rainbow trout LC50=5540 mg/L/96H Sunfish (tap water), death at 14250 ppm/24H Mosquito fish (turbid water) TLm=13000 ppm/48H
Environmental Fate:
   Volatilizes, leeches, and biodegrades when released to soil.
Physical/Chemical:
   No information available.
Other:
   None.

Appendix I

**** SECTION 13 - DISPOSAL CONSIDERATIONS ****

Dispose of in a manner consistent with federal, state, and local regulations.
RCRA D-Series Maximum Concentration of Contaminants: Not listed.
RCRA D-Series Chronic Toxicity Reference Levels: Not listed.
RCRA F-Series: Not listed.
RCRA P-Series: Not listed.
RCRA U-Series: waste number U002 (Ignitable waste)
This material is banned from land disposal according to RCRA.

**** SECTION 14 - TRANSPORT INFORMATION ****

  US DOT
    Shipping Name: ACETONE
    Hazard Class: 3
      UN Number: UN1090
    Packing Group: II
  IMO
    Shipping Name: ACETONE
    Hazard Class: 3.1
      UN Number: 1090
    Packing Group: 2
  IATA
    Shipping Name: ACETONE
    Hazard Class: 3
      UN Number: 1090
    Packing Group: 2
  RID/ADR
    Shipping Name: ACETONE
  Dangerous Goods Code: 3(3B)
    UN Number: 1090
  Canadian TDG
    Shipping Name: ACETONE
    Hazard Class: 3
    UN Number: UN1090
  Other Information: FLASHPOINT -20 C

**** SECTION 15 - REGULATORY INFORMATION ****

A. Federal
  TSCA
    CAS# 67-64-1 is listed on the TSCA inventory.
  Health & Safety Reporting List
    None of the chemicals are on the Health & Safety Reporting List.
  Chemical Test Rules
    None of the chemicals in this product are under a Chemical Test Rule.
  Section 12b
    None of the chemicals are listed under TSCA Section 12b.
  TSCA Significant New Use Rule
    None of the chemicals in this material have a SNUR under TSCA.
  SARA

Material Safety Data Sheet

Section 302 (RQ)
  None of the chemicals in this material have an RQ.
Section 302 (TPQ)
  None of the chemicals in this product have a TPQ.
SARA Codes
  CAS # 67-64-1: acute, chronic, flammable, sudden release of pressure.
Section 313
  This chemical is not at a high enough concentration to be reportable under Section 313.
  No chemicals are reportable under Section 313.
Clean Air Act:
  This material does not contain any hazardous air pollutants.
  This material does not contain any Class 1 Ozone depletors.
  This material does not contain any Class 2 Ozone depletors.
Clean Water Act:
  None of the chemicals in this product are listed as Hazardous Substances under the CWA.
  None of the chemicals in this product are listed as Priority Pollutants under the CWA.
  None of the chemicals in this product are listed as Toxic Pollutants under the CWA.
OSHA:
  None of the chemicals in this product are considered highly hazardous by OSHA.
B. State
  2-propanone can be found on the following state right to know lists:
  California, New Jersey, Florida, Pennsylvania, Minnesota, Massachusetts.
  California No Significant Risk Level:
  None of the chemicals in this product are listed.
  MSDS Creation Date: 11/30/1994 Revision Date: 8/13/1996

The information above is believed to be accurate and represents the best information currently available to us. However, we make no warranty of merchantability or any other warranty, express or implied, with respect to such information, and we assume no liability resulting from its use. Users should make their own investigations to determine the suitability of the information for their particular purposes. In no way shall Fisher be liable for any claims, losses, or damages of any third party or for lost profits or any special, indirect, incidental, consequential or exemplary damages, howsoever arising, even if Fisher has been advised of the possibility of such damages.

Tables

Vapor Pressure of Liquid Water at Various Temperatures

Temperature (°C)	Vapor Pressure (torr)	Temperature (°C)	Vapor Pressure (torr)
10.0	9.209	26.0	25.209
12.0	10.518	27.0	26.739
13.0	11.231	28.0	28.349
14.0	11.987	29.0	30.043
15.0	12.788	30.0	31.824
16.0	13.634	31.0	33.695
17.0	14.530	32.0	35.663
18.0	15.477	33.0	37.729
19.0	16.477	34.0	39.898
20.0	17.535	35.0	42.175
21.0	18.650	40.0	55.324
22.0	19.827	45.0	71.88
23.0	21.068	50.0	92.51
24.0	22.377	55.0	118.04
25.0	23.756	60.0	149.38

Density of Water at Various Temperatures

Temperature (°C)	Density (g/mL)	Temperature (°C)	Density (g/mL)
10.0	0.9997026	26.0	0.9967867
12.0	0.9995004	27.0	0.9965162
13.0	0.9993801	28.0	0.9962365
14.0	0.9992474	29.0	0.9959478
15.0	0.9991026	30.0	0.9956502
16.0	0.9989460	31.0	0.9953440
17.0	0.9987779	32.0	0.9950292
18.0	0.9985986	33.0	0.9947060
19.0	0.9984082	34.0	0.9943745
20.0	0.9982071	35.0	0.9940349
21.0	0.9979955	40.0	0.9922187
22.0	0.9977735	45.0	0.9902162
23.0	0.9975415	50.0	0.9880393
24.0	0.9972995	55.0	0.9856982
25.0	0.9970479	60.0	0.9832018

Appendix J

Molarities of Concentrated Acids and Bases

Solution	Concentration
Ammonia, NH$_3$	15 M
Hydrochloric Acid, HCl	12 M
Nitric Acid, HNO$_3$	16 M
Sulfuric Acid, H$_2$SO$_4$	18 M
Phosphoric Acid, H$_3$PO$_4$	18 M

Solubility Rules for Ionic Compounds

Substance	Rules
Na$^+$, K$^+$, NH$_4^+$, NO$_3^-$	Ionic compounds which contain these ions are soluble.
SO$_4^{2-}$	Most ionic compounds containing sulfate are soluble. Exceptions are: BaSO$_4$, SrSO$_4$, PbSO$_4$, CaSO$_4$, Hg$_2$SO$_4$, and Ag$_2$SO$_4$ (partially soluble).
Cl$^-$, Br$^-$, I$^-$	Most compounds containing chloride, bromide, or iodide are soluble. Exceptions are: AgX, Hg$_2$X$_2$, PbX$_2$, HgBr$_2$, and HgI$_2$. PbCl$_2$ is somewhat insoluble.
Ag$^+$	Ionic compounds containing Ag$^+$ are insoluble except AgNO$_3$.
OH$^-$	Most ionic hydroxide compounds are insoluble. Exceptions are: NaOH, KOH, NH$_3$(aq), Ba(OH)$_2$ (somewhat insoluble), and Ca(OH)$_2$ (partially soluble).
S^{2-}	Most ionic sulfide compounds are insoluble. Exceptions are: Group IA, Group IIA, and NH$_4^+$ sulfides.
CrO$_4^{2-}$	Most ionic chromate compounds are insoluble. Exceptions are: K$^+$, Na$^+$, NH$_4^+$, Mg$^+$, Ca^{2+}, Al^{3+}, and Ni^{2+}.
CO$_3^{2-}$, PO$_4^{3-}$, SO$_3^{2-}$, SiO$_3^{2-}$	Most ionic carbonate, phosphate, sulfite, and silicate compounds are insoluble. Exceptions are: K$^+$, Na$^+$, and NH$_4^+$ compounds.